W9-ATF-220

People and Pollution

People and Pollution

Cultural Constructions and Social Action in Egypt

Nicholas S. Hopkins
Sohair R. Mehanna
Salah El-Haggar

The American University in Cairo Press
Cairo New York

113 Sharia Kasr el Aini, Cairo, Egypt
420 Fifth Avenue, New York, NY 10018
http://www.aucpress.com

Dar el Kutub No. 4677/01
ISBN 977 424 663 2

Photographs by
Salah El-Haggar
Muhammad Hassan
Nicholas S. Hopkins
Hugh Hughes
Said Samir

Designed by the AUC Press Design Center/Hugh Hughes
Printed in Egypt

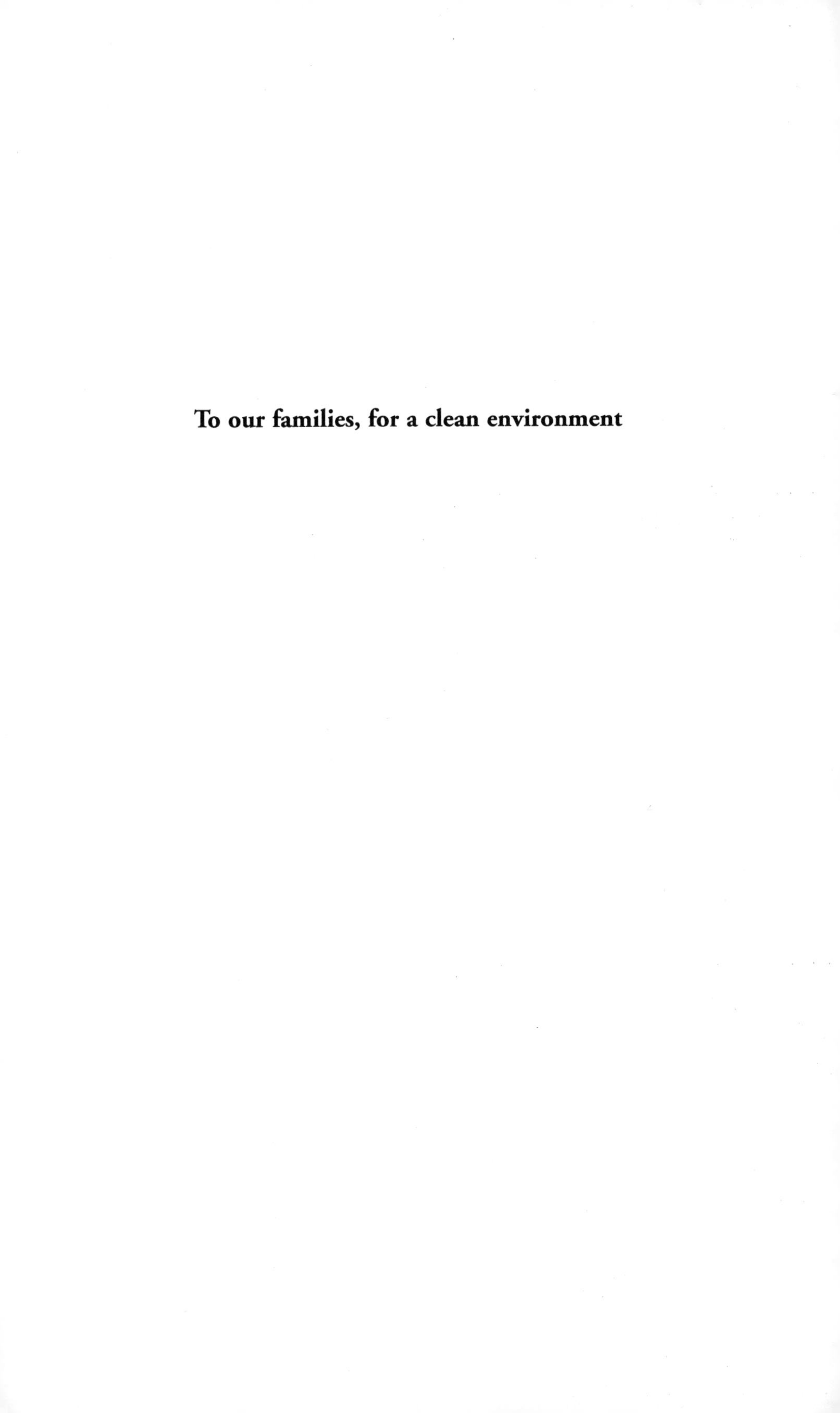

To our families, for a clean environment

Contents

Tables

Illustrations

Foreword

In an increasingly complex and interdependent world, there is less tolerance for polluting attitudes and actions toward our fragile environment and its natural resources. Ever since 1994, when Law 4 for the Protection of the Environment was ratified, environmental issues have been high on Egypt's political and public agenda. The vision of the Ministry of State for Environmental Affairs can be summarized in one sentence: we work for the present with an eye on the future. Prudent management of our natural resources and strategic environmental planning are essential ingredients for sustainable social and economic development. Our aim in Egypt is to integrate an environmental dimension into the very fabric of society and the challenge is to use an integrated approach that strikes a sound balance between environmental protection and economic and industrial growth. To implement this vision, we are bringing together complementary expertise and skills from Egypt and the international community. We are actively involving governmental and non-governmental stakeholders across Egypt with the view that constructive dialog and effective coordination are critical to realizing our environmental objectives. While there has been notable progress in overcoming the problems created by an accumulation of more than fifty years of pollution, there is yet much to do and it requires everyone's cooperation.

It therefore gives me great pleasure to contribute a foreword to this study of the reaction of people in Egypt to environmental pollution and degradation, as well as to the efforts to sustain and preserve our natural resources. This work, by a team of distinguished social scientists and engineers from the American University in Cairo, and carried out with Canadian support, is one of the first works anywhere in the world to examine how citizens feel about the environmental issues around them, and how they find it possible to address these problems, many of which cannot be "wished" away.

This original study is based on extensive field research with both academic and policy relevance. The uniqueness of the book comes from its focus: instead of the usual approach of analyzing policy and measurements, this text seeks to understand how the people themselves, often the objects of policy, understand their environment and their own actions. What is evident is that there is a popular Egyptian environmentalism that with support and encouragement could take many of these problems in hand. This is clearly a response both to broader values in Egyptian culture and to the circumstances where people live.

People and Pollution is an impressive and useful initiative to be consulted by environmentally concerned readers, students, and policymakers alike. I hope it will inspire and mobilize its readers to take pragmatic action, since action always speaks louder than words.

Nadia Makram Ebeid
Minister of State for Environmental Affairs

Preface

Our research project on the social response to environmental change and pollution in Egypt was inspired by the growing environmental problems of Egypt and the rising level of concern about these visible—both at the level of government and at the level of individual Egyptians. There has been a veritable explosion of research and writing on environmental issues around the world in the last decades, but relatively little of that has focused on rigorous analysis of the feelings and experiences of people and communities, especially in urban areas. Filling this gap is one of the main contributions this study hopes to make. Reflecting people's life experiences, this is one type of "Egyptian environmentalism."

Our study shows the kind of thinking about environment and pollution issues that is found among nonelite, nonexpert, "ordinary" Egyptians. The level of concern is high, and partly overlaps with the concerns of experts. The main areas of concern are garbage, sewage, and clean streets, followed by air, water, and noise pollution. People sometimes take matters into their own hands, especially when a problem can be resolved by cooperation among neighbors, though they are disinclined to confront major polluters. People tend to place the most responsibility for pollution on others like themselves. Global issues such as the depletion of the ozone layer, the protection of nature, or global warming tendencies are not much on people's minds. The major metaphor is "cleanliness" and its opposite, "dirtiness," rather than "nature," and the focus is on the "brown agenda" rather than the "green agenda." Methodologically, we focused on cultural and social processes in the community rather than individual perceptions.

Our study was funded by the International Development Research Centre (IDRC), headquartered in Ottawa, Canada. The research was carried out by a team based at the Social Research Center (SRC) of the American University in Cairo (AUC). The research period began in January 1995, and continued through the summer of 1997, with the initial drafting of this report extending until 1998.

The 1994 proposal to the IDRC followed a successful application, in collaboration with Princeton University, to the John D. and Catherine T. MacArthur Foundation for seed money to prepare the project. This pilot project enabled us to address some of the issues involved in doing research on environment and pollution in Egypt, and to carry out exploratory field research. This research was carried out in Cairo and Kafr al-Dawwar (Beheira governorate) in the summer of 1993, and a summary of the results was published as "Pollution and people in Cairo" (Hopkins et al., 1995). Many of the quotes at the heads of chapters come from this study.

Our interest in this project took shape when the general level of concern for environmental issues at official and public levels was rather low; we have seen this concern grow during the course of our project. One sign of that growing concern was the establishment of the Ministry of State for Environmental Affairs in 1997, after the bulk of our research was done. This ministry incorporated the Egyptian Environmental Affairs Agency (EEAA), which assisted us in the early stages of our research.

Our project was of necessity a team project. It involved, first of all, collaboration between the two social scientists, Nicholas S. Hopkins and Sohair Mehanna, coordinated through the SRC at AUC. A team of engineers led by Dr. Salah al-Haggar and Dr. Samia Galal of the AUC Department of Engineering collaborated with the social scientists, so this study is also an outcome of cooperation between social science and natural science. In addition to social scientists and engineers, the team included resident researchers, enumerators, coders, translators, and others.

The drafting of this report was also a collaborative enterprise. The first drafts of the different chapters were written variously by the social scientists, the engineers, and the resident researchers. This final version was then prepared by the two senior social scientists, though the others were always prepared to answer inquiries in their areas of expertise.

The resident researchers were key participants in our team. Without their intelligence and loyalty, the research would never have been completed. They were individually responsible or shared responsibility for each of the research localities. They included Muhammad Abdelhakim, Tamer Abdelkader, and Ibrahim Hassan (all Kafr al-Elow), Gamal Abdelaziz and Said Samir (Dar al-Salam), Muhammad Hassan (Sayyida Zeinab), and Hayam Ahmed (Abkhas).

In the early phases of our research, two anthropology students wrote master's theses on issues related to the environment. Inas Tawfik and Eman el-Ramly also contributed to the various discussions and conferences that we had then and later. In a broader sense, we consider them part of this project, and thus their work is cited with gratitude in this book.

Other SRC staff members also helped. As usual, Fikri Abdelwahhab and Esmat Kheir were diligent in making contacts, keeping the project running smoothly, and in organizing the two dissemination seminars. Hanan Sabea provided irreplaceable assistance in organizing and leading the focus groups. Ekbal Sameh provided expert help with computers and word processing.

Suzanne Hammad and Elisabetta al-Karimy were efficient research assistants at different points in the research, and able translation assistance was offered by Mona al-Adly, Dalia Hamed, Dalia al-Bayoumi, and Dalia Wahdan. The list of enumerators and coders is given below.

Walton Chan prepared the maps with intelligence and dispatch, while Eric Denis at CEDEJ (Cairo) provided some of the basic data.

We are grateful to various colleagues who helped us in the early stages of the research program. From AUC's science department, we benefited from the advice of Dr. Jehane Ragai, Dr. Fadel Assabghy, and Dr. Ashraf al-Fiqi. During the pilot project, we collaborated productively with Dr. John Waterbury and Dr. Stephen Brechin of Princeton University. Dr. Russell Stone of the American University in Washington DC provided opportune comments. Dr. Sahar Tawila of the SRC advised us on sampling methods. Dr. Salwa Gomaa, now of Cairo University, contributed to the early formulation of the problem.

From both an administrative and a scientific point of view, we enjoyed the support of Dr. Hoda Rashad, the director of the SRC, and her predecessor, Dr. Saad Nagi. Effective support was also provided by Dr. James Collum and Mouna Shaker of AUC's Office of Sponsored Programs, and their staff. Many others at AUC and in the wider "environment" community in Cairo provided moral support during the long research process. Special thanks to those who in various ways maintained our computers.

This report could not exist without the cooperation of thousands of Egyptians in the four research localities and elsewhere, who made an effort to answer our questions forthrightly. We hope that the information contained in this report will contribute to an amelioration of the conditions of life for ordinary Egyptians, and so that the effort of all will not have been in vain.

Names of team members (resident researchers)
Muhammad Abdelhakim
Tamer Abdelkader
Ibrahim Hassan
Gamal Abdelaziz
Said Samir
Muhammad Hassan
Hayam Ahmed

Enumerators and coders
Abu Bakr Taha
Ahmed Hamed

Ashraf Abdel Fattah
Hala Ismail
Hana Mohamed Fadl
Hassan Abdel Sattar
Hossam Abou al-Fetouh
Huwayda Hakim
Loay Ahmed Mohamed
Mohamed Attiyatalla
Nagla Fathi
Raghda Abdel Fattah
Samer Fuad
Sayed Ali al-Rayis
Sherifa Abdel Fattah
Wael Adel Nabi
Yasir Fayez Abdel Zaher
Zeinab Mohamed Hafez

Map 1: Greater Cairo and the lower Delta

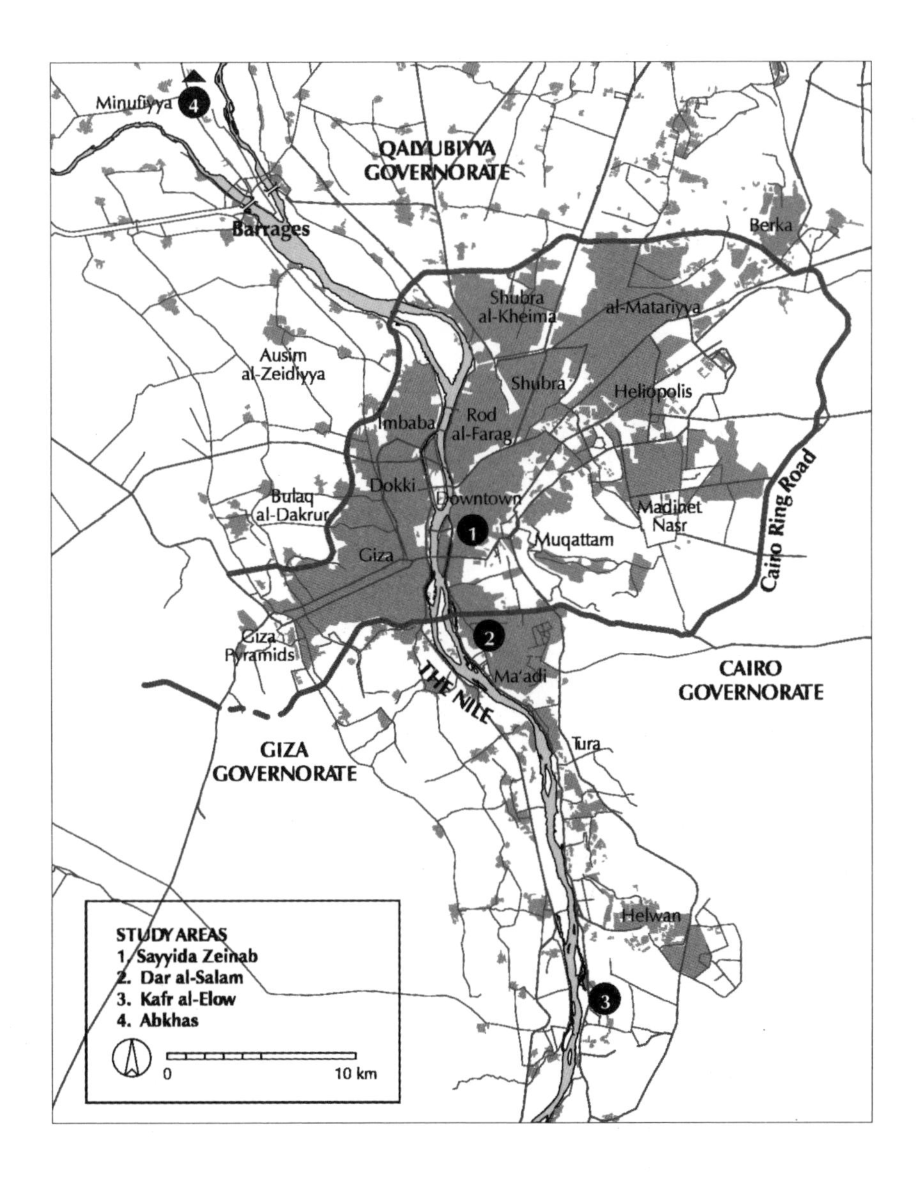

Map 2: Kafr al-Elow

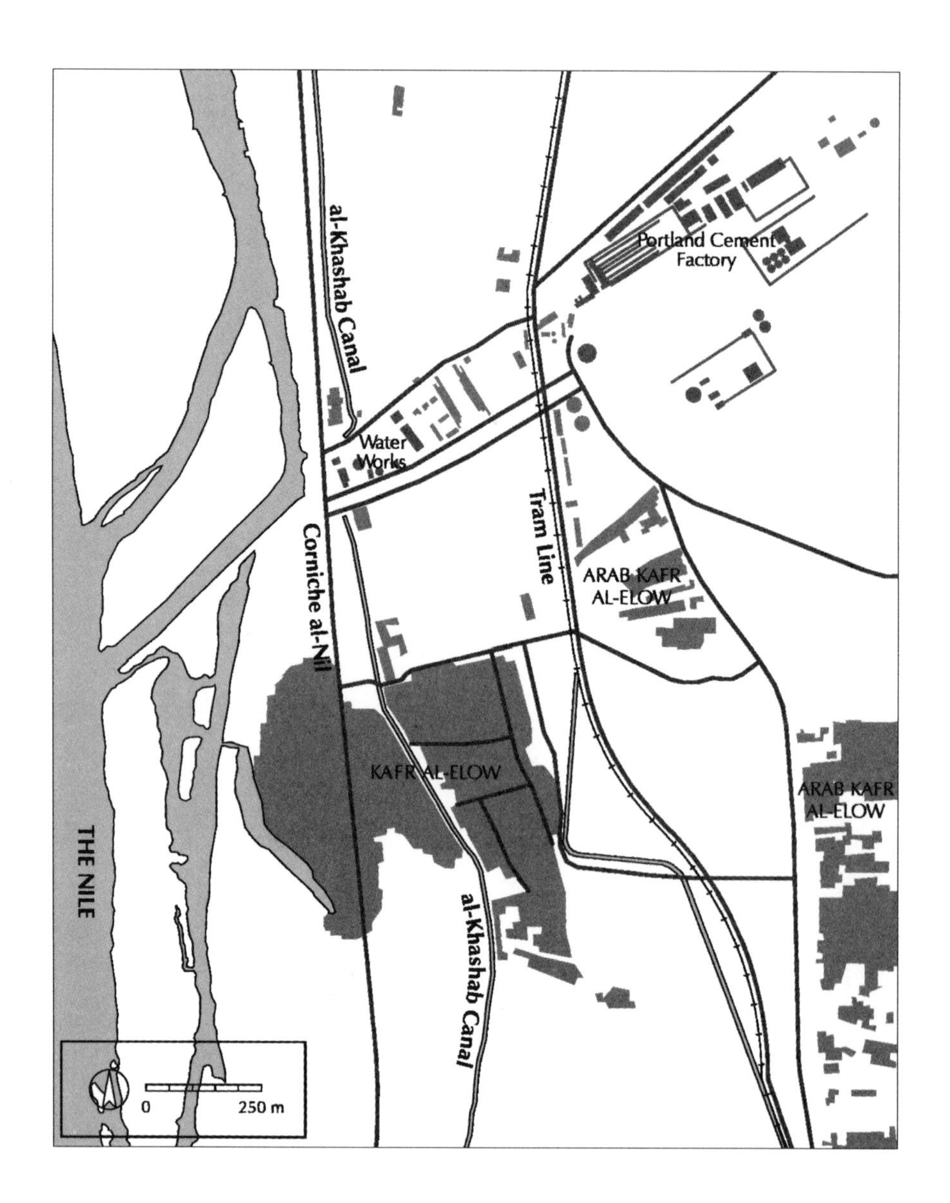

Map 3: Dar al-Salam

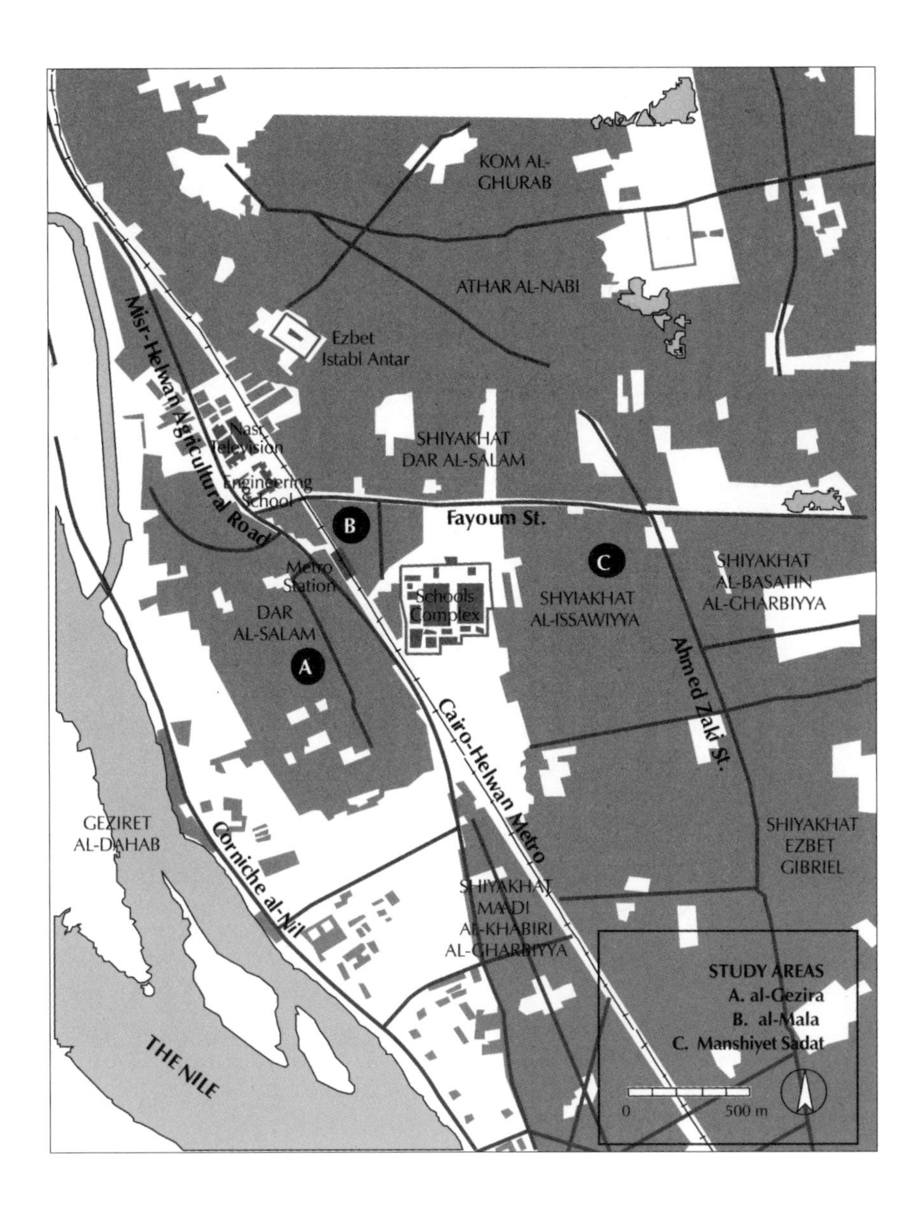

Map 4: Sayyida Zeinab

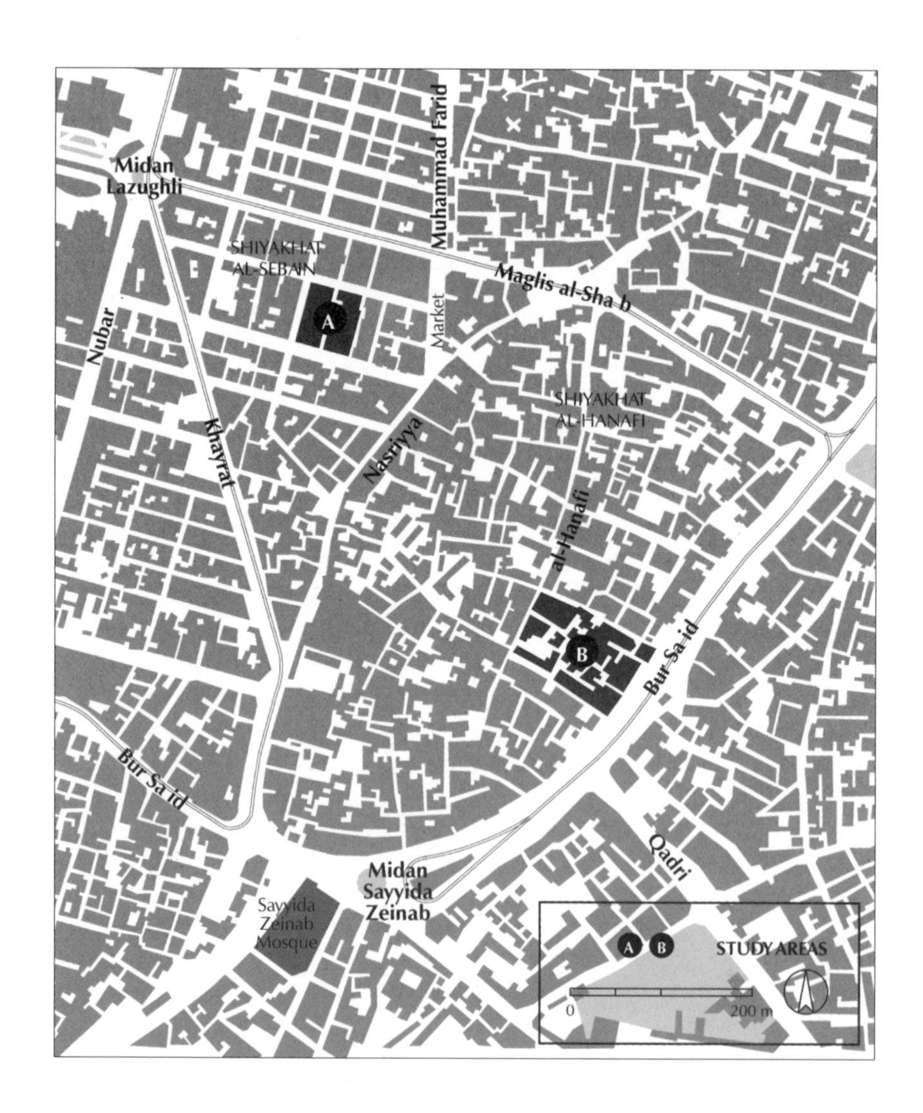

Map 5: Abkhas

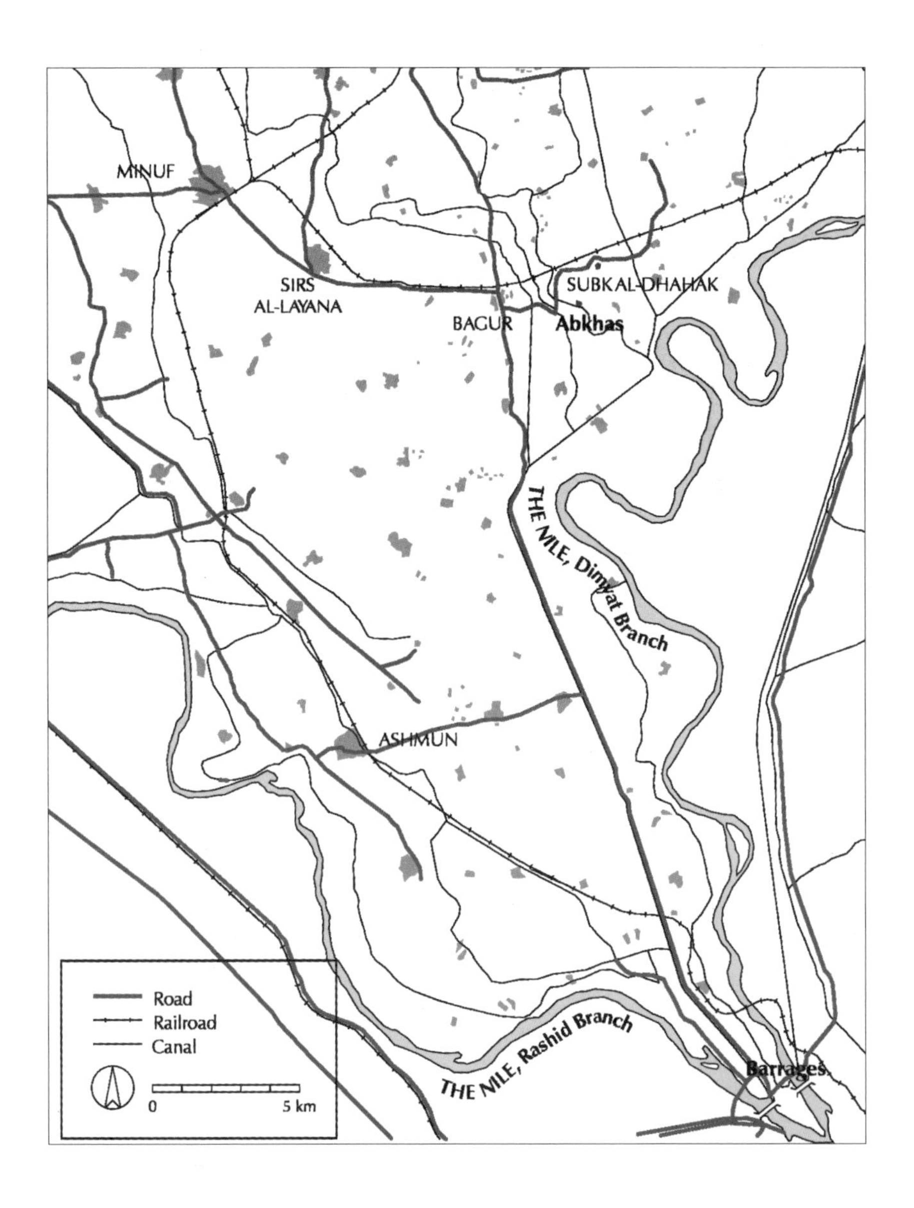
MINUF
SIRS
AL-LAYANA
BAGUR
SUBK AL-DHAHAK
Abkhas
THE NILE, Dimyat Branch
ASHMUN
THE NILE, Rashid Branch
Barrages
Road
Railroad
Canal
0
5 km

1

Environmental Change and Social Response in Egypt

> We are in a crowded area. So you see, the water, air, and food are of bad quality in our area. We have factories that emit smoke that is poisonous, we have bad quality water, the streets are not paved, and garbage is everywhere. So you see, we are a polluted community.
>
> Taxi driver, Shubra al-Kheima, 1993

The Research Problem

People living in working-class and middle-class areas of urban and rural Egypt are concerned about environmental issues and pollution, and are aware of many of the factors involved in understanding these problems and in acting with regard to them. Some 95% of our respondents were able to identify at least one environmental problem in their neighborhood. This awareness, which many observers have assumed did not exist, is the subject of this book. Thus the issue is not whether there is an environmental awareness in Egypt, but what the nature of that awareness is, how it varies from one setting to another, and above all where it leads. That awareness identifies environmental problems as primarily related to health and to cleanliness, rather than to nature as in the West. Efforts to maintain a clean and healthy environment flow from this, but people must struggle against adverse physical circumstances and what they see as an indifferent government.

The recent development of Egypt has led to such advances as an increase in the literacy rate, in life expectancy, in income levels, and in infrastructure construction, and to a decrease in infant mortality and in the rate of population growth. But as with development elsewhere in the world, environmental problems have arisen. These include problems with air and water pollution, with noise, with the spread of agricultural chemicals, with the loss of farmland and virgin coastline, and with the disposal of solid and liquid human wastes. Some of these are concomitant with the industrialization of Egypt, much of which occurred at a time when environmental considerations were routinely ignored around the world. Others are related to the increase in motor vehicles, or to shifting technology, such as the changing use of fertilizers and pesticides in agriculture, or simply to increasing human density. Although elites continue to make most of the critical decisions concerning the environment, the middle and lower classes of the population are increasingly feeling the effects of these decisions.

Our book reports the results of a research project on the social and cultural dimensions of environmental change and the response to that change. Environmental change includes environmental degradation in the form of pollution and the emergence of environmental hazards. The response to that change derives from culturally constructed understandings and perceptions of the change as well as the social actions made possible by these cultural constructions. We prefer the concept of "cultural constructions" to "perceptions" or "attitudes" in order to focus attention on interaction within a community or group rather than on the isolated individual. The constructions are cultural (content oriented) but also social (process oriented). Our assumption was that grassroots actions (to ameliorate or change a situation) spring from understandings that are produced in interactions among people. Our research was carried out in four localities in Egypt, described in the next chapter.

We propose a model: detrimental environmental change (pollution or degradation) leads to both physical and social stress, which is understood through a process of cultural construction of reality; this cultural construction in turn frames and shapes the actions of individuals, particularly as they are motivated by a sense of risk (typically to health); and these actions may remain as the acts of individuals or may evolve into social movements or other forms of change under specific circumstances.

Environmental Stress and Human Input in Egypt

According to the 1996 census, Egypt had a population of 61.4 million, of whom 2.2 million were abroad. This was an increase of 22% over the 1986 population. The rate of population growth was 2.1% annually, at that time meaning an additional 1.3 million people per year. The population was 57% rural and 43% urban, though larger rural communities were often quite urbanized. The population of Greater Cairo alone, according to the census, was 11.5 million, of whom about 25%–30% lived in so-called "informal zones," or self-built slums (Ismail 1996:121).

Environmental stress in Egypt is caused by air pollution from car exhaust, open-air burning, and factories, by water and soil pollution (especially in rural areas) from chemical runoff, by noise pollution linked to overcrowding, by poverty-linked problems such as poor disposal systems for solid and liquid wastes, by insects and rodents, and by the spread of such diseases as schistosomiasis due to changing irrigation patterns (Watts and El Katsha 1995, 1997). Egyptians often also mention "moral pollution." An often-cited study of Cairo problems carried out for USAID (PRIDE 1994) cites the three highest health risks from environmental sources as particulate matter, lead (in all media), and microbiological diseases from environmental causes. These factors are compounded in the Egyptian case because of the concentration and density of the population (Said 2000). Some of these issues are relatively "democratic" in their implications (air pollution from automobile exhaust), while others affect the poor more heavily than the rich and thus pose equity problems (low-income housing adjacent to industrial zones, overcrowding, poor disposal of solid and liquid wastes).

Egypt's problems reflect human decision-making, and the resolution of these problems in turn depends on human organization and agency. Urban Egypt faces problems similar to those found in other third world cities, most notoriously the mushrooming cities of south and southeast Asia, but also in Africa and Latin America (see WHO/UNEP 1992), while rural Egypt shares the problems of intensely cultivated areas everywhere. There is a rich comparative field here that we can barely evoke: do the different cultural and political settings account for the fact that environmental movements emerge more prominently and effectively in some countries than in others? Here we can only examine this question in the context of Egypt, leaving the comparison for another day.

Green and Brown Environmental Problems in Egypt

Our focus in this study is on the so-called "brown agenda"—"the problems of pollution, poverty, and environmental hazards in cities" (World Resources

Institute 1996:19), extended to include similar problems in rural Egypt. Specific issues include water and sanitation, housing, food contamination, pests, indoor air pollution, ambient air pollution, solid and hazardous wastes, lead exposure, occupational exposure, traffic accidents, urban ozone, industrial pollutants, and air and noise pollution due to motor vehicles. The World Resources Institute (WRI) report notes (1996:19):

> Although most of the world's population will soon be living in developing world cities, the environmental problems most prominent in these cities have often been conspicuously absent from the global environment agenda. Indeed, over the past two decades the global agenda has shifted away from local and regional problems such as air pollution and inadequate water supplies toward vast global concerns such as ozone depletion, climate change, and the loss of biological diversity.

Clean water and air, as well as clean streets and a minimum level of quiet, are not only pleasant but clearly essential to health. In many third world cities today the absence of these features contributes to the sense of poor health and well-being that people have—especially poor people living in marginal conditions (Briscoe 1993; Campbell 1989; Hardoy et al. 1990; Sengupta 1999).

The "brown agenda" problems are not the only environmental issues in Egypt. Others include the adequacy and reliability of the water supply of Egypt (Said 1993), the protection of the Red Sea coral reefs and of marine and desert wildlife in general (biodiversity) (Hobbs 1995), the rapacious development of the coastlines in the past 20 years (Cole and Altorki 1998), and the historical preservation of the built environment, including ancient monuments. Many of these issues are important for tourism. In addition, Egypt is potentially affected by such global problems as ozone depletion, the loss of biological diversity, biosafety issues, and climate change. It has, for instance, been suggested that global warming could result in a decrease of as much as 25% in the runoff of the Nile (Engelman and LeRoy 1993:34), not to mention the flooding of a large part of the Delta if sea levels rise.

Government Action

In 1992, the Egyptian government summed up its environmental policy in these words: "Egypt adopts an environmental policy that supports sustainable development programs taking environmental considerations in perspective and provides a life fit for its citizens."[1] The three pillars of environmental policy in Egypt are the institutional framework, the legal framework, and the National Environmental Action Plan (NEAP) (Gomaa 1997:61). From 1982 to 1997 the main Egyptian government institution concerned with the environment was the Egyptian Environmental Affairs Agency (EEAA); in 1997 this agency was absorbed into the new Ministry of State for Environmental Affairs (MSEA). In early 1994, the Egyptian People's Assembly passed an environmental law, Law 4 of 1994, that provoked vigorous debate not only in the Assembly, but also between ministries, and with the involvement of nongovernmental organizations (NGOs) (Gomaa 1997:53). Gomaa summarizes the law (1997:41):

> The law motivates polluters to follow a more sound environmental behavior by setting standards, requiring environmental impact assessments, allowing for an adjustment period for industrial establishments, forcing them to comply with regulations, and offering economic incentives for the first time in Egyptian law. Penalties are called for in the law as a last resort against those polluters who are violating the rules. The law also stipulates the responsibilities of the Environmental Affairs Agency in environmental management and creates a special fund to support environmental activities.

A year later, the implementing regulations were issued, allowing three years for factories and other polluters to bring themselves into compliance with the law's provisions. Thus, March 1, 1998 was set as the date by which factories had to clean up their emissions, though many factories did not achieve this target. The effectiveness of this law in reducing industrial pollution in Egypt has not yet been determined, but the effort continues.

The Egyptian government has been extending the sewage network in Cairo, maintains a reasonably good water-purification system (though the quality of the distribution pipes and storage tanks is another question), is trying to improve the supply of clean drinking water in the village, is trying to control factory emissions, and recently substituted methyl tertiary butyl ether (MTBE)[2] for lead in the gasoline. The government often appears not to be responding to grassroots pressure, but rather to the opinions of Egyptian and international experts (see Gomaa 1997:33, 62).

The Egyptian government's determination to clean up the environment is steadily increasing, and there is a correspondingly higher level of awareness on the part of people. Thus, our analysis has a moving target. Nevertheless, we were often surprised by how ignorant official actors (including bilateral and multilateral agencies, NGOs, etc.) were of grassroots thought and practice. There is much more concern at an official level with the technology of environmentalism than with its human element.[3] Conversely, the level of concern at the grassroots level is not matched by the quality of information available to the public.

Seminars and workshops on the impact of environmental change in Egypt invariably come around to the need for "public awareness" or "participation." For instance, at the celebration of Earth Day, April 24, 1993, at AUC, a distinguished Egyptian scientist, Mustafa Tolba, the long-time director of the United Nations Environment Program (UNEP), spoke of the need to involve the public in sustainable development, and went on to refer to the "skill and wisdom of local people" (Tolba 1993). Egyptian activists often stress the need for more voluntary associations to spread this awareness or mobilize people. Yet little is known of how these "local people" perceive the changes in their environment, nor of what action they may be prepared to take in response. The assumption is that "the people" are an empty slate on which leaders are free to inscribe what they wish.

There are, broadly speaking, two paths of action for dealing with environmental threats. One path is through government action: the establishment and enforcement of regulations, investment in water-purification plants, sewage systems, "cleaner" industrial machinery, and the like. In some cases, this implies government action against corporate polluters; in others, government regulation of itself. The other path is through changes in the behavior of individual citizens. At some point, people become convinced that there is a "risk" from the environment, and then, depending on the cultural resources available to them, they respond. These responses may be either individual or collective, and they may or may not lead to government action. Various NGOs in Egypt have tried to mobilize people around a sense of risk (Gomaa 1992; see also Ibrahim et al. 1997), but without much success. On the other hand, there may be effective local organizations, or even individuals, whose actions are all the more successful for not being widely known. For instance, individuals or small groups may seek relief from pollution by contacting individual officials or politicians, and may be effective in obtaining relief, albeit limited (Tewfiq 1996). There are reports of street blocks organizing to maintain cleanliness (Hopkins et al. 1995; el-Ramly 1996). And some environmental NGOs have been successful at working at the grassroots level, such as the Association for the Protection of the Environment that works with the *zabbalin,* or garbage collectors, (Assaad and Garas 1994) or the Arab Office for Youth and Environment, which has organized cleanups in various parts of Cairo (Gheith 2000).

The best known example of social mobilization around pollution issues comes from the part of al-Waily, in northern Cairo, known as Ezbet Mekawy (Kamel 1994; Tewfiq 1997). A pair of lead smelters raised severe environmental concerns among residents of this area, which includes densely populated low-income residential areas and industrial zones. An alliance formed between local residents and an outside NGO oriented primarily to health issues to force these smelters to close or move. Over the years, a variety of techniques were used, including complaints to officials, publicity in newspapers and on radio, petitions, the gathering of scientific evidence on the high lead content of children's blood, and research into the legal situation of the smelters. A key role was played by local members of parliament, one of whom belonged to the Tajammuʿ opposition party. In fact, residents with a common left-wing or Nasserist background were significant players in the local organizing network. Other techniques, including demonstrations and appeals to non-Egyptian NGOs and aid agencies, were also contemplated but rejected as too risky. The goal was to make an overwhelming case for the closure of these two smelters. One of the doctors involved wrote (Kamel 1994:33):

> Many residents were eager to fight against the lead smelters but expressed fear of arbitrary arrest or some other action by the government for participating in the group's activities. For some, this feeling of powerlessness was expressed by calls for violent action. One resident made a statement that echoed the feelings of many: "If we were in a war against the factory owners, they would be considered war criminals for using chemical weapons against women and children. The only solution is to destroy the factory by planting a bomb."

Meanwhile, the owners of the smelters used counter-tactics—there were allegations of bribery, countersuits, claims that the number of employees was so high that closure would be very disruptive, efforts to hide information, false certificates of compliance, playing different branches of the government off against each other. Eventually, in 1994, the smelters were closed down and remain closed. However, other industries remain in the area.

Theoretical and Conceptual Approaches

We consider here four theoretical approaches to understanding the relation between a people and the environment in which they live. The first approach is symbolic and focuses on the cultural construction of environmental issues. Notions of cleanliness, for instance, create moral communities and establish boundaries. The second is a form of risk analysis, and basically asks when and how a sense of risk produces an action, and what kind of action. Is that action, for instance, individual or collective, such as a social movement? The third is derived from the seminal article of Garrett Hardin on "the tragedy of the commons" (1968), the notion that a clean environment is a "commons," or common property; Hardin's argument seeks to contrast the self-interest of individual behavior threatening this commons to the ability of people organizing in defense of it. And the fourth derives from the "environmental justice" paradigm and analyzes how environmental problems may be understood in terms of relations among people rather than between people and environment. Environmental change may affect groups differentially, and thus the cultural constructions can be interpreted also within a class framework.

The Social and Cultural Construction of Environmental Issues

An abundant literature now stresses that environmental issues are "'constructed' by individuals or organizations who define pollution or some other objective condition as worrisome and seek to do something about it" (Hannigan 1995:2). Hannigan adds that "social constructionism focuses on the social, political, and cultural processes by which environmental conditions are defined as being unacceptably risky and therefore actionable [requiring action] . . . . A social constructionist approach . . . recognizes the extent to which environmental problems and solutions are end-products of a dynamic social process of definition, negotiation, and legitimation both in public and private settings" (pp. 30–31). Stressing its complexity, Hannigan has defined cultural construction as "a framework of contested definitions, frames, and meanings" (p. 190).

A major question has been the extent to which these constructions reflect direct observation of environmental circumstances and changes on the one hand, or the content of the mass media and the opinions of specialists on the other. From this point of view, the research question centers on the origin of people's understanding and how these sources of knowledge are turned into cultural constructions. Hannigan stresses "the central role of media discourse and that of science in interpreting and shaping the contexts, conditions, and consequences of the environmental crisis" (pp. 56–57), rather than a sense of threat from the observed environment. In considering why only some environmental claims capture public attention and become politically relevant, it is important to analyze the role of the media and other public debates.

A further question relates to the link between these cultural constructions, or frames within which intentions to act can be formulated (Čapek 1993:7), and political action (Hopkins and Mehanna 1997, 2000). Many studies from around the world have focused on environmental movements. Čapek herself analyzed the process whereby a neighborhood of African-Americans became aware of the environmental hazards in their residences and sought redress, using techniques and symbols drawn from the civil rights movement. Levine (1982) and Fowlkes and Miller (1987) have analyzed the Love Canal episode, in which a neighborhood in New York State mobilized, with media support, to secure redress from the toxic environment in which they lived. Guha (1997) highlights the role of outside organizers utilizing a "vocabulary of protest" in organizing a movement in rural India to defend local resources from corporate takeover from the outside.[4] In all of these cases, there is a mixture of internal recognition of the problem, public recognition and perhaps definition through the media, and a role for outside sympathizers. These cases show a tendency for locally inspired movements to become national or international in scope.

Two major recent studies of the cultural construction of environmental situations come from France and the U.S. These two studies illustrate different approaches in the effort to construct cultural models of the environment for given populations and attempt to evaluate what this means in terms of action. Both studies start from the notion that environmental change occurs and is the initial impetus to the elaboration of cultural constructs. Both conclude that environmental action is often less than one might expect given the amount of thought and concern about environmental issues. Those who are concerned locally lack the skills to mobilize in their own defense.

France

Françoise Zonabend (1993) has written an account of reactions to the presence of a nuclear plant at La Hague in rural France, both by the neighbors of the plant and by the workers in it. There is a threat to health from radioactive exposure from both the work at the plant and the waste it produces, a threat that can be all the more worrisome because in the short run it is invisible. Zonabend's focus is on people's speech, on what they say and do not say, what they hint at through metaphors and constructed silences, the imagistic language they use to describe the presence of a nuclear risk. "The spoken word, in this context, becomes a vehicle for any number of ruses designed to obscure the ostensible, purported meaning of the narrative heard. Language may tell or leave untold, guide or mislead, shed light or spread confusion" (Zonabend 1993:3). People "borrow . . . from the collective memory of the group;" they use images drawn from the cultural constructions of their past to grasp and articulate the present issues of contamination and irradiation (p. 125). Zonabend's argument is that everyone in this context feels anxiety, though they may have different ways of expressing it. "The sense of risk and the fear of danger exist in everyone, regardless of gender, social and occupational background, or professed opinion" (p. 123).

The question then arises, what sort of action results from this set of cultural notions? Some of the people Zonabend describes succeeded in denying or suppressing their concern: Zonabend even speaks of a "pact of silence" and a certain "fatalism" of the people of La Hague, when faced with the overwhelming weight of the government, the elite, the French people in general: small populations feel obligated to go along. An equally common reaction by some individuals confronted with a perilous situation is to take more chances, out of a sort of machismo. Such people are often those who know most, or think they know, in this case exemplified by workers in the nuclear plant. However, an effective countervoice to the plant gradually emerged, in large part due to people outside the local community, growing louder especially after Chernobyl in 1986 (pp. 63–68). During the late 1990s the international environmental organization Greenpeace orchestrated much of the opposition to the nuclear plant at La Hague.

United States

Kempton, Boster, and Hartley (1995) conducted a highly original research project into U.S. environmental values, basing themselves on cognitive and cultural anthropology. Working from data on individual mental models, they attempted to construct a cultural model of U.S. environmentalism within the framework of American culture. ("Our direct data are the *mental models* of the individuals we interview, which, when we find them widely shared, we argue are American *cultural models* (p. 11, emphasis in original). The establishment of such a model then allowed them to analyze the patterns of variation away from that model, since it is accepted that no cultural model will be shared by all individuals who might be considered part of that culture. The model contains existential elements, reflecting what people think the world is really like, and value elements, reflecting a sense of the desirable. One source of variation is between those with specialized, perhaps scientific knowledge, and those who are participants in the general cultural setting without such knowledge.

Kempton, Boster, and Hartley find that there is a dominant American set of environmental beliefs and values, ascertainable from roughly three quarters of their respondents. They contrast this consensus to issues like abortion or gun control, where there appear to be two alternative and contrasting viewpoints (p. 211). The implication of this relative consensus (or paradigm) is that it may be easier to mobilize citizens around courses of action that (may) require sacrifice since they share a starting point. They sum up by saying (pp. 215–16) that in the U.S.

> lay environmentalism is built upon cultural models of how nature works and how humanity interacts with it, and is motivated by environmental values. Cultural models of the way nature works include the complex interdependencies among species and other systems, human reliance on the environment, and the way human activities affect nature. Environmental values include humanity's utilitarian need for nature, obligations to our descendants, the spiritual or religious value of nature, and for some, the rights of nature in and of itself. Finally, American environmentalism represents a consensus view, its major tenets are held by large majorities, and it is not opposed on its own terms by any alternative coherent belief system.

This model or paradigm is consistent with core American values, such as parental responsibility, obligation to descendants, and traditional religious teachings, and includes the notion of valuing nature for its own sake ("biocentrism") (p. 214). Kempton et al. argue that policy makers in the U.S. should not only stress utilitarian values but also refer to religious teachings and biocentrism to make their cases (p. 224).

Once analysis yields the model, one can estimate how people will act, or what kinds of policies they will support, even if at some cost to themselves in other areas. Kempton et al. stress the concept of appropriate action: the question they ask is whether the cultural models lead people to appropriate action, defined as action that considered scientific opinion supports (p. 218). Nevertheless, they also raise the broader question of why there is not more environmental action, given the cultural model. The answer they find is that sometimes people choose inappropriate actions because the cultural model makes erroneous assumptions, and that sometimes there are external constraints to action. Thus, they conclude, "for environmentally beneficial actions, environmental beliefs and values are necessary but often are not sufficient, given the multiple existing barriers to action" (p. 220).

Although both studies work from a notion of cultural construction, one can note that Zonabend argues that speech only partly and indirectly reveals people's feelings and understandings, while Kempton, Boster, and Hartley take speech as indicative in itself. Moreover, Kempton, Boster, and Hartley argue that the cultural construction can be judged against scientific opinion, while Zonabend implicitly treats it as a critique of the official scientific policy of encouraging the use of nuclear power.

Our first goal in this study is to examine the nature of the cultural construction of the environment and pollution in Cairo and in an Egyptian village, keeping in mind Hannigan's warning that cultural constructions are contested domains. We will see that the cultural construction of the environment that dominates in Egypt focuses on threats to health and on the related issue of cleanliness, rather than on relations to nature, and that much importance is given to the possible role of the government. The dominant metaphor in people's discourse is "pollution," related to dirt and health, and not "shortage," related, for instance, to resources.

Risk and Action

Ulrich Beck has argued that the world is moving towards a "risk society," in which the central political conflicts are not class struggles over the distribution of money and resources but instead non-class-based struggles over the distribution of environmental and technological risk (Beck 1992:111). Beck refers to "risk winners and risk losers" (p. 112). Beck mostly has in mind the threat from nuclear accidents, for instance, where the risks can no longer be calculated the way an insurance risk can be, and where the criterion is the "worst imaginable accident."[5] Nuclear contamination is egalitarian and so "democratic": high social class protected no one from contamination after Chernobyl. Most forms of air and water pollution are comparable. For Beck, the solution is more public involvement in decisions affecting technology—more democracy. He notes that "the question . . . of how the universal challenge of an industrial system producing wealth *and* destruction is to be solved democratically remains completely open, both theoretically and practically" (p. 117; emphasis in original). He concludes this article with the piquant observation that if radioactivity itched, and so were perceptible, we might do more about it!

The identification, prioritizing, and ranking of risk are all social processes, founded on a certain construction of risk, and with implications for behavior in response to the risk. The American geographer, Gilbert White, notes that "unless a risk analysis comprehends the social structure within which individual decisions are made, it may fall far short of understanding either the process or the consequences of those decisions" (White 1988:174). Numerous studies have shown that for perception of change or even deterioration to turn into a perception of risk requires factors other than the scientifically ascertainable level of risk, factors that identify, amplify, or attenuate the risk (Kasperson et 1988). "Whether or not something is perceived as hazardous may be related at least as much to social factors as to quantified estimates of risk" (Wolfe 1988:4).[6] Douglas and Wildavsky note that "the perception of risk is a social process . . . . The different social principles that guide behavior affect the judgment of what dangers should be most feared, what risks are worth taking, and who should be allowed to take them" (1982:6), and continue, "Risk taking and risk aversion, shared confidence and shared fears, are part of the dialogue on how best to organize social relations" (1982:8).

The process through which these judgments form is a social or collective one in that it involves relations among people, but the basis of the judgment is a cultural one, for it includes values and a sense of the good as well as what Kottak (1999:26) calls an "ethnoecology"—a cultural model of the environment and its relation to people and society. Douglas and Wildavsky argue that "only a cultural approach can integrate moral judgments about how to live with empirical judgments about what the world is like" (1982:10). Thus, they note, "pollution ideas are the project of an ongoing political debate about the ideal society" (p. 36), and are eventually "an instrument of control" (p. 47). Dirt, as Douglas famously argued in another context (1970 [1966]:48), is "matter out of place"; the notion of "out of place" combines a moral and a physical judgment, and thus points us to the existence of symbolic boundaries between (imagined) groups.[7] Her argument also reminds us that the opposite of dirt is order, in other words, that a clean environment is an orderly one, with all the moral connotations that implies. "Dirt offends against order. Eliminating it is not a negative movement, but a positive effort to organize the environment" (p. 12).

So perceptions of risk by people in general often do not match those of the experts (see Kempton et al. 1995), who also may not agree among themselves. These perceptions are cultural "models of," to use Geertz's terminology (Geertz 1973); they are descriptive. But they are also prescriptive, and each "ethnoecology" implies a certain course of action (Kottak 1999:29). It becomes, in Geertz's terms, a "model for" action. One question therefore is how and when does a sense of risk based on a perception of environmental pollution evolve into a course of action, and what is the nature of that action: is it collective (political action to accompany the public debate) or is it adaptive behavior by individuals?

There is thus a useful sense in which acts can be seen as occurring within a framework of perceptions, cultural constructions—in a word, "thought."[8] How do people cope with a risk or a hazard that they have come to recognize? Theoretical and comparative literature links social movements to environmental questions (see Levine 1982, Johnson and Covello 1987, Newman 1992, Čapek 1993). This is one way. However these actions may also be individual, ranging from avoidance (moving to a cleaner neighborhood) to political (individual appeals to authorities to solve a problem). And of course, "denial" is always a possibility.[9] But the first step in understanding these activities is to understand the cultural template that guides them. The role of culture here should be seen as enabling rather than determining, in line with current anthropological thinking on the issue. The perception of risk alone is not adequate: people must have the means and the power to respond to the threat.

The prioritization of risk is implicit in people's behavior, more perhaps than in their words. People make choices among water sources, try to protect themselves from particulate matter in the air or insects in their houses, hire others to do the dirtiest jobs, etc. There are gender differences, based on conventional understandings of male and female tasks, as well as on the different kinds of health risks that people take or find supportable. People interpret many ailments and problems as "natural," therefore not worth seeking care for. For example, women in Egypt are sometimes reluctant to seek health care because of their responsibilities in the family and because they fear to appear to their husbands as a "sickly wife" (Khattab 1992). On the other hand, they are responsible for the health care of their children, and are alert in this area. Thus their sense of risk varies, and with it their response.

Much of the cultural construction of risk and of the environment can be linked to the general economic and social conditions in which people live. For the poor, the choice between environment and development is acute, and the options are fewer than for the privileged. Poverty entails economic pressure to accept risk. The poor often articulate an opposition between economic development, and hence jobs, in a competitive world, on the one hand, and environmental cleanliness on the other, not knowing how to achieve both. Nowadays it happens that most people earn their living in the paracapitalist world of contemporary Egypt—they work for a salary, or are small-scale entrepreneurs in the private sector. Even in the village, less than half the men in our sample claimed to be farmers, and even farmers are entrepreneurial commodity producers. In both village and city, women are also directly involved in this world, not simply through being housewives/home managers relying on their husbands' incomes from salary or profit. The sharpest conflict was felt by men and their families when the men work in hazardous environments such as cement plants or lead smelters. Although very few people said they would choose to work in a hazardous environment (some actually do, of course), even fewer said they would allow their child to work in one.

People do not react to the risks the experts identify, but to those they feel themselves, or have learned to feel. Hence, our second goal in this study is to identify the risks that Egyptians feel and to evaluate what the consequences of this construction are for action.

The Property Rights Approach—the Degradation of the Commons

In his classic article on "the tragedy of the commons," Garrett Hardin devoted a few lines to the link between pollution and the "commons," or common property resources (1968:1245):

> In problems of pollution . . . it is not a question of taking something out of the commons, but of putting something in—sewage, or chemical, radioactive, and heat wastes into water; noxious and dangerous fumes into the air; and distracting and unpleasant advertising signs into the line of sight. The calculations of utility are much the same as before. The rational man finds that his share of the costs of the wastes he discharges into the commons is less than the cost of purifying his wastes before releasing them. Since this is true for everyone, we are locked into a system of "fouling our own nest," so long as we behave only as independent, rational, free-enterprisers.

In other words, a clean environment shares some of the features of a commons: all stakeholders have an interest in a clean environment, but as individuals some abuse that commons. Hardin's solution for this conundrum was the use of "coercive laws and taxing devices that make it cheaper for the polluter to treat his pollutants than to discharge them untreated,"[10] since the water and air cannot be privatized in an effort to keep them free from pollution. He did not consider the strength of local social organization to achieve the same end.

Recent literature on the commons, drawing primarily on anthropological studies of productive commons, stresses that "societies have the capacity to construct and enforce rules and norms that constrain the behavior of individuals . . . . A diversity of societies in the past and present have independently devised, maintained, or adapted communal arrangements to manage common-property resources . . . these arrangements build on knowledge of the resource and cultural norms that have evolved and been tested over time"

(Feeny et al. 1990:13).[11] In this tradition, common-property resources are defined as "a class of resources for which exclusion is difficult and joint use involves subtractability" (p. 4). In our case, waste must be discarded so in the long run exclusion is virtually impossible, and careless disposal of waste subtracts from the cleanliness of the environment, thus reducing its utility to others. These authors also distinguish between the situation of "open access," where there are no well-defined property rights and access to the resource is unregulated and free and open to everyone, on the one hand, and "communal property," where collective ownership is recognized and rules apply, on the other. "Open access" describes the situation of water and air in Egypt.

Our third goal here is to examine these issues with regard to the situation in Cairo, extending the logic of collective action and the commons to a clean environment. We examine the extent to which there are communal or collective arrangements to manage the resource of cleanliness by constraining the behavior of individuals.

Environmental Justice and Social Movements

The final and fundamental question of the study is to investigate how environmental problems variously affect different elements of Egyptian society (by residence, class, gender, etc.), and to what extent this differential impact is a part of the construction that Egyptians put on the situation. We can analyze the situation according to the environmental justice paradigm, and we can follow how perceptions of injustice (as contrasted with perceptions of threat) can lead to social action or social movements. We can ask, with Putnam (1995), what are the cultural resources (cultural capital) that people can bring to bear on issues of collective action, and what are the structural or other constraints that impede such efforts.

The notion of environmental justice refers to the situation in which the decisions of the wealthy and powerful create environmental problems for the relatively poor, while leaving the wealthy relatively unbothered. Thus, the focus is on the relationships of groups with each other, rather than with the physical environment. Examples include the siting of factories or hazardous waste sites in poor neighborhoods, or the dumping of hazardous waste in places or even countries that cannot protect themselves. And in some cases, the victims of this situation rise up and attempt to protect themselves against polluters. One of the early U.S. cases in which environmental justice was an issue was a proposal to establish a potentially toxic landfill in a poor, predominantly African-American county in North Carolina (Bullard 1990; McGurty 1997); the residents of this county, black and white, formed themselves into a social movement to fight off this threat. Similarly, researchers examined all 415 hazardous waste facilities in the U.S. in 1987 and determined that "people of color were about twice as likely as white people to live in the towns that hosted such facilities" (Sachs 1995:10). Many of the problems of pollution discussed in our study of Egypt pose equity issues that affect the poor more than the rich, children more than adults, certain neighborhoods more than others, even certain countries more than others. People must not only come to grips with the pollution problem, but also with the implications of that problem for their position vis-à-vis others in society.

When a danger or risk from environmental situations threatens some categories of the population more than others, social movements often emerge.[12] The empirical question in Egypt is to see whether this happens, and under what conditions. However, before a social movement appears, a goal has to be identified; in other words, the responsible parties must be identified. And the shape of the social movement will reflect both the group that creates it and the definition of the problem (the "frame") that this group holds. Who, in other words, is to blame? It is important to know whether people essentially blame themselves,[13] or whether they blame a power outside themselves and their community. To what sources or agents (e.g., government, business, their neighbors, poverty, themselves) do the different groups and categories attribute responsibility for the degradation of their environment?

Social movements also appear when middle class or elite individuals, specialists, and others set out to correct an environmental problem. Many writers trace the origins of the environmental movement to the U.S. publication of Rachel Carson's *Silent Spring* in 1962, which led to movements opposing the extensive use of pesticides and fertilizers in agriculture (Bell 1998:173–74). The international environmental movement, Greenpeace, originated in Canadian efforts in 1969–1971 to protest U.S. nuclear testing in the Aleutian Islands of Alaska (Wapner 1996:44–46), and the Danish environmental movement took a major move forward during the (successful) struggle in the 1970s to dissuade the Danish government from establishing nuclear power plants (Jamison et al. 1990:90–109). These movements all started from a cultural construction of risk (from pesticides or nuclear radiation, or from destruction of wilderness), but they also all started as movements of a few skilled and sophisticated individuals, able to use the political system to affect policy.

A large part of our investigation in Egypt was devoted to identifying social actions or movements, or, alternatively, the reasons why people might appear passive in the face of threats and injustice (as was sometimes true in the French and U.S. cases cited above). The national-level environmental movement in Egypt is weak, and so many of the actions we found were partial and tentative. On the other hand, the potential for action from the state or from the elite is tremendous, and for this action to be effective, it must be realistic about the situation on the ground. Egyptians are torn between cynicism and hope: hope that powerful actors can be persuaded to take an interest; pessimism that they actually will, for long.

Conclusion: Lessons from the 'Black Cloud'

These different points (construction of a problem, common property, risk and action, and environmental justice) can be illustrated by the "black cloud" episode in Cairo in October 1999.[14] This was a severe episode of air pollution that caused concern on the part of many articulate citizens. It was certainly not the first such episode, but it was the first time that people "noticed" it. In this story, we can see how "awareness" of pollution arose, as it were for the first time, among people of all classes, and how they attempted to understand it, in a sense to agree on what construction to put on this reality. It was generally agreed that the problem arose because of the inconsiderate acts of some who were polluting the common clean air. However, although all classes suffered from the greater air pollution, only the articulate classes could make themselves heard, and they, through the media, tended to blame the less fortunate for the problem. In other words, the problem was seen as one of relation between two groups or classes rather than between humans and nature—the environmental justice paradigm. This cultural construction in turned framed the actions of individuals, so the sense of risk to health turned into pressure on the government to "do something." The government in turn had to show that it was responsive, although in the end it did not convince everyone that it was effective in dealing with the "crisis."[15]

On the evening of October 23, 1999, many people in Cairo noticed particularly heavy air pollution. This came to be known as the black cloud of pollution. This "cloud" reduced visibility, and caused respiratory problems and eye irritation. Many people called the police and fire stations, and also the newspapers and the EEAA to express alarm and to discover what was happening. "The smoke has panicked the city's 16 million inhabitants, who demanded an explanation, especially after various authorities gave conflicting interpretations for its cause" (*Egyptian Gazette* 1999). Over the next couple of weeks there were numerous stories and much commentary from leading columnists in the press, reporting complaints from those serious citizens who could pick up the telephone and call government officials and newspaper columnists. Many of the commentators did not find the government's efforts to account for the black cloud convincing.[16] With perhaps some exaggeration, Salama Ahmed Salama called the pollution an "environmental catastrophe." Salah Montasser said he received 82 telephone calls complaining of distress (*al-Ahram*, Nov. 1, 1999). One of his correspondents complained that when he tried to find out from the Environment Ministry what the cause was, he was put off by being told to submit a written complaint.

The first statements on the cause of the smog came from the weather bureau, which blamed the air pollution on smoke from burning rice straw in the Delta combined with a thermal inversion. On the strength of this, several Delta governors banned the burning of rice straw. Many others did not find this explanation convincing. Farmers were quoted in the press as saying that the amount of rice straw they burn could not be the cause of this pollution in far-away Cairo. The Ministry of Agriculture requires farmers to burn rice straw and cotton stalks to prevent the spread of insects, so the farmers were caught between two stones. There were also some suggestions that the ash from rice straw is used as a fertilizer for the next crop, and on the other hand, some expert opinion suggested that burning rice straw could indeed produce harmful fumes.

Two days into the crisis, the Minister of State for Environmental Affairs borrowed a helicopter from the Minister of Defense to conduct a survey of possible sources, and reported that burning garbage, industrial waste, and automobile exhaust appeared to be the main sources. Meanwhile, the Minister of Information mentioned "factories, foundries, and car fumes." The head of the Cairo Sanitation and Beautification Authority denied, implausibly, that there could be any burning garbage in Cairo. More fancifully, some rumors identified other possible sources such as an explosion in a chemical factory, a nuclear leak, the ongoing Egyptian-American military maneuvers, or even the change to a new prime minister. The source of the air pollution remained undetermined, and the press often held the government responsible for not pinning down the source. There was no report in the press of any effort to analyze the chemical content of the air to identify the source, though Dr. Magdi Allam of the EEAA did admit that the particulates in the air were much higher than average.[17] Dr. Mahmoud Al-Heweihi of the Institute of Ecological Studies and Research at 'Ain Shams University noted that the symptoms people were complaining of were more likely to be caused by burning garbage, which produces sulfuric oxide, nitrogen, and hydrocarbonates, than burning rice straw, which yields carbon dioxide.

The publicity meant that the government had to do something. As it happened, the cabinet was brand-new, having taken office earlier in October.

The first action was the helicopter inspection by the Minister of State for Environmental Affairs. A week after the first experience, October 30, 1999, it was reported that "Prime Minister Ebeid met with ministers of agriculture, information, environmental affairs, interior, transport, and local development as well as the governors of Cairo, Giza, Qaliouba, and Sharqia and the chairman of the Egyptian Meteorological Authority" (*Egyptian Gazette* 1999). This meeting seemed to place most of the blame on burning agricultural waste and garbage, since the outcome was to allocate LE275 million to further garbage recycling. Later it was announced that a fleet of trucks would haul away the rice straw so that farmers would not have to burn it.

A scientific committee, headed by the Minister of Higher Education and Scientific Research, was set up, and "concluded that the smog is the result of certain meteorological conditions that were exacerbated by the presence of smoke in the air—from the burning of agricultural refuse, car exhaust fumes, and factory emissions. When the air contains a particularly powerful cocktail of pollutants and there is no wind for consecutive days, smog appears" (*Al-Ahram Weekly* 1999:3). The phenomenon seems comparable to the photochemical "brown-air smog" (Miller 1998:466).

The event made people in Egypt more aware than previously of the threats from environmental pollution. Pressure on the government, largely through newspapers, meant that a more public response was required. It also meant that somewhat stricter measures to control pollution could be undertaken. The Minister of State for Environmental Affairs took advantage of the situation to shut down about 200 small polluting establishments around Cairo, and a major effort to haul away accumulated garbage from Cairo was undertaken. (She also canceled a trip to a conference on climate change in Germany, symbolizing the power of the local over the global.) Discussion continued intensely for several weeks, then over the next several months there were references to the fall's "black cloud" as something to be avoided.

This air pollution crisis in October 1999 represented a growing awareness among the elite of the seriousness of the problem, and also illustrated some of the mechanisms available for dealing with it—individual complaints, the role of the press and of respected columnists, dramatic but short-term government action, the weight of elite pressure in the debate. It also illustrated the tendency of the elite to find that the poor (rice farmers and urban dwellers without garbage collection) were responsible. As Douglas and Wildavsky have pointed out (1982:37), concerns about environmental pollution are often used to reinforce social boundaries, or to see environmental problems as problems of relations among people. Although newspaper accounts often mentioned automobile or factory emissions among the possible sources, these points were not mentioned as often as open-air burning of waste by the poor. Finally, the episode illustrated some of the tension and ambiguity in the dyadic relationship between government and people, as people complained of feeling left alone by the absence of government action and explanation.[18]

This book is a study of a growing environmental awareness in Egypt, with special attention given to the ordinary citizens who live in middle class and lower class areas in Cairo and the countryside. In it we examine the levels of pollution that can be found in our study localities, and the construction of that reality by residents. We look at the conditions under which people live, and the steps that they feel able to take (or perhaps the ones they avoid taking) in order to create a picture of changing environmental circumstances in Egypt, and of the reaction or response to this pattern of change.

2

Methodology

> We Egyptians are always complaining. We are always feeling that we are in the worst possible situation, whether things are improving or not.
>
> Student, Imbaba, 1993

Our book is primarily based on the research project we carried out over a three-year period, from early 1995 to mid-1997. Our approach was as social scientists, specifically guided by an anthropological frame of reference. Our first step was to identify reasonably bounded social units where we could concentrate our efforts. Within these research areas or localities, we then conducted surveys, organized focus groups, interviewed key informants, followed local events, and measured air and water quality and noise levels. Our methodological goal was to situate the numerical results coming from a survey in the context of neighborhood social and cultural data coming from intensive contact with a single locality. We refer to those who answered the survey questions as "respondents" and those who shared in the focus groups as "participants." While social science data necessarily come from individuals in the first instance, our concern was not with a collection of individual "perceptions," but with the collective "construction." Thus, we fit broadly within the constructivist paradigm, with a goal of elaborating a cultural model, as the discussion of theoretical approaches in Chapter 1 already indicates.

Research Questions

In the previous chapter we spelled out our theoretical and conceptual concerns, and our research questions follow directly from them:

- What is the scope and nature of environmental degradation in the four sites?
- What social and cultural constructions of environmental degradation in these sites do those who reside and work in these areas share? And through what mechanisms are they constructed?
- How are environmental perceptions affected by a range of background factors such as sex, age, class, occupation, or place of residence?
- What is the role of the mass media in influencing perceptions of environmental degradation?
- To what sources or agents (e.g., government, business, neighbors, poverty, informants themselves) do the different groups and categories of individuals attribute responsibility for the degradation they see in their environment?
- How and when does this understanding lead to individual or collective action?

We need to enter a couple of disclaimers. Except when clearly indicated otherwise, we are relying on what our respondents and participants say, and so that is where the truth value lies, not in the content. In particular we do not have the resources to verify the allegations of pollution we heard, so we take them as perceptions or constructions. Our data are what people say, and our analysis is of this discourse and its origins. Our concern as social scientists is not with the problems themselves but with how the people contacted through our research program saw the problems, though as human beings and responsible citizens, we are also interested in the "real" problems. Moreover, although our data are not representative in a technical sense, we believe they generally reflect Egypt as a whole. From time to time we have referred to "Egyptians," which should be understood in the strict sense to implicate only those Egyptians who talked to us, unless a broader reference is clearly intended.

Research Sites

We selected four sites for our study, three in Cairo and one in the rural Delta. We sought sites that were diverse in their problems and histories, and confirmed this through preliminary visits. In the end we selected an urban area notoriously affected by pollution, an urban area not situated close to any visibly polluting industry but with problems of solid waste disposal and populated by the working class, an urban area with more of a middle class population, and a rural area with a diversified agriculture but without a distinctive source of pollution. These sites are described below.

For the Cairo sites, we started by choosing the community of Kafr al-Elow to represent the heavily polluted industrial zone of Helwan. We considered similar areas in northern Cairo, but Kafr al-Elow has another feature that prompted our choice: the anthropologist Hani Fakhouri studied the community under the auspices of the SRC in the 1960s. From that experience we have a record of an earlier phase in the community's history (see Fakhouri 1987), and we retain some personal contacts that facilitated our start-up.

Once we chose Kafr al-Elow, we then looked for two other sites that would represent different situations and would not be too widely scattered geographically. These sites would be "ordinary" in the sense of having no outstanding local source of pollution. After examination of some alternatives, we chose an older inner city area, the *shiyakhas* (wards) of Saba'in and Hanafi in the *qism* (district) of Sayyida Zeinab, and a newer, informal area, Dar al-Salam. Dar al-Salam is sprawling, and internal political boundaries are lacking or unclear, so we ended up interviewing in three sublocalities that reflect three of the major subdivisions of the area—al-Gezira, al-Mal'a, and Manshiyet Sadat, each of which is newer and more prosperous than the preceding.

We also hoped to extend our survey to Ma'adi, to expand the range of classes covered, but this did not work out. The upper middle class and upper class residents of Ma'adi are much more reluctant to be interviewed than the residents of the other neighborhoods where we worked. Even focus groups were hard to schedule. In the end we include some information on Ma'adi, but it is much less systematic than for the other sites. This is drawn from the excellent study by Eman al-Ramly (1996) and from some focus group material.

We examined several villages in Qalyubiyya and Minufiyya governorates before settling on the village of Abkhas, in the Bagur district of Minufiyya governorate. We were looking for a village that would be sufficiently small that the whole population would be our sample, and one that did not have some distinctive and perhaps peculiar geographic feature. For instance, one proposed village was on the banks of the Nile, and another was bisected by a major road. We sought a village that would be far enough from Cairo to be away from major urban pollution effects, yet close enough for us to reach easily. Our choice of Abkhas represented the convergence of these various factors.

Kafr al-Elow

Kafr al-Elow is a former village that has been partly transformed into a working-class neighborhood for workers from the nearby factories in the Helwan industrial zone.[19] Kafr al-Elow is part of the Helwan area that is presented in the Egyptian press as the most polluted area in greater Cairo. Environmental problems include industrial waste from nearby factories, poor sewage and solid waste disposal,[20] flies and mosquitoes, and noise and other problems related to the heavy trucking along the main road.

Situated on the Corniche road, Kafr al-Elow is located southwest of Helwan and north of Tibin. The 1996 census recorded a population of 38,565 in the *shiyakha* of Kafr al-Elow, which is slightly larger than the zone where we focused. This is up 39% from the 27,778 recorded in 1986. Although urban and even industrial in its occupational structure, in its social organization Kafr al-Elow retains some village features such as the prominence of powerful extended families. These families maintain guest houses, provide charity, and dominate politics.

The main features of the locality are the old Khashab canal, the Corniche highway, and the nearby factories of which the Helwan Portland Cement Co. is the most prominent. Also in the area are many industrial establishments, such as another cement company (the Qawmiyya Cement Co.), steel works, a coke factory, a textile plant, a cardboard box factory, foundries, and others. Many small workshops, especially truck repair shops, line the highway. Kafr al-Elow was formerly a farming community, but little farmland remains. There are schools, mosques, private doctors and pharmacies, a police station, and a telephone exchange.

Dar al-Salam

Dar al-Salam is our example of a recent informal (*'ashwa'iyya*) area. We recognized four sublocalities within this complex locality: al-Gezira, al-Mal'a, Manshiyet al-Sadat, and Old Dar al-Salam. We did the survey and the focus groups in the first three of these, of which al-Gezira is the poorest and Manshiyet al-Sadat the newest and best off. Most residential areas are occupied by multistory buildings (from three to seven floors for the most part) built along narrow streets with poor air circulation. There is little planned empty space.[21]

In the 1996 census, the district of al-Basatin, which includes our research zone, had a population of 666,928 (up 48% from 449,556 in 1986), while 'Isawiyya ward, which included most of our research areas, had a population of 93,380 (up 26% from 74,380 in 1986, and 161% from 35,710 in 1976) and adjacent Dar al-Salam ward had a population of 96,844 (up 18% from

81,825 in 1986, 272% over 26,038 in 1976). Dar al-Salam grew up after 1960 between downtown Cairo and Ma'adi (Abu-Lughod 1971:201–02; see also El-Kadi 1987, Kharoufi 1991, 1995), and our figures show that the main growth in the area occurred between 1976 and 1986. Since the area is recent, most people were born elsewhere and moved in. This is reflected in the pattern of voting registration, as people are frequently registered in their original homes.

Most apartments have electricity and running water. Waste water in many areas was still being collected in lined tanks,[22] but a new sewage system was under construction during our research. There are schools, private doctors, and some other services in the area, but the police station is small, there is no fire station, and the administrative headquarters is distant. The area has been served by the Metro (Cairo lightrail service) since 1988, and microbus routes link the Metro station to the various neighborhoods. A number of voluntary associations are registered in Dar al-Salam of which about 10 are active, mostly charitable organizations with a religious orientation.

Dar al-Salam is probably the most polluted of our sites despite the reputation of Kafr al-Elow (see below). The main sources of pollution in Dar al-Salam are sewage problems, uncollected garbage that is sometimes burned, air and noise pollution from traffic on the main streets; dust from unpaved streets, refuse and noise from workshops, raising poultry at home and in the street, and general overcrowdedness. Noise comes from workshops, cassette players, children playing, itinerant vendors calling out their wares, and the use of loudspeakers for weddings, funerals, the call to prayer, or advertising. People fret over insects and rodents. There is some unease about social relations, as evidenced by people's concern about "newcomers" with different values or about moral pollution, for instance the behavior of boys toward girls in the streets or the use of drugs and *bango* (marijuana).

Sayyida Zeinab

Sayyida Zeinab is our example of an older, inner city habitat. The inhabitants are generally from the middle and lower class. The district is centered on the famous mosque and shrine devoted to Sayyida Zeinab (Abu-Zahra 1997). Our focus was on two adjoining wards called al-Hanafi and al-Saba'in in the north of the district. Al-Hanafi includes some mosques from the Mamluk period, while al-Saba'in was built mostly after 1870 (Arnaud 1991). In the 1996 census, Sayyida Zeinab district had a population of 156,142, whereas al-Hanafi recorded 15,048 and al-Saba'in 7,136. As in other inner city areas of Cairo, the population is shrinking. Al-Hanafi and al-Saba'in together had 34,843 people in 1976 and 22,184 20 years later (a drop of 36%), while

Sayyida Zeinab as a whole lost 22% of its population between 1986 and 1996. Younger families are moving out to newer areas such as Dar al-Salam. The area contains a mixture of old and new buildings. Services and utilities are generally present in this old, well-established area. Schools and medical care are readily available, and transportation is not a problem.

The main problems are air pollution and noise. There are the usual problems with poor garbage collection, and there are some industries (notably sweet factories) interspersed among residential buildings. A daily street market discards a lot of organic waste. An obsolescent sewer system was being rebuilt during our research. Inhabitants also complain of moral pollution, because of the presence of drugs and cafés that show dubious videos. In general cafés serve as social centers and as venues for election campaigns. The National Democratic Party is fairly active in the area, and some of its members take an interest in urban environmental issues.

Abkhas

Abkhas is located on both banks of the main irrigation canal for the central Delta, the Bahr Shebin. It is part of the administrative village of Subk al-Dhahak, in turn part of Bagur *markaz* (district) in Minufiyya governorate. Abkhas is about 4 km east of Bagur, just south of the paved road from Bagur to Benha, down an unpaved lane of about half a kilometer.

The population of Abkhas was 2,919 in the 1986 census and 3,344 in the 1996 census, an increase of 15%. The population is less than half that of an average Egyptian village. In addition to the main village there are four attached hamlets named Tajammu' al-Kubri, 'Ezbet al-Bahr, 'Ezbet Qandil, and 'Ezbet Abul-Enein. The village *'omda* is assisted by two *shaykh al-balad*s and two *shaykh al-'ezba*s. The total land area is 548 feddans (230 ha), of which buildings cover 45 feddans (19 ha) and public utilities 64 feddans (27 ha), leaving 439 feddans (184 ha) for agriculture. There are no large landholders in the village; the average holding is 1.5 feddans (0.6 ha), close to the Minufiyya average (Ireton 1998:50). The main crops are wheat and clover in the winter and maize and sweet potato in the summer, and people keep numerous domestic animals.

Abkhas has two primary schools, three mosques in the main village and one in each hamlet, shops, and a cooperative. Healthcare is available in neighboring villages. Electricity is available, but water is not reliable. People get water from three sources, none of them good: the government supply, piped in from tanks in the main village of Subk al-Dhahak, and ultimately from a deep well; pumps, reaching ground water at 20m to 30m; and canal water. The preferred drinking water is from the pumps. Village houses have unlined latrines. There is no solid waste collection system, and some waste is dumped into or near the canal.

The main environmental problems come from the use of chemicals in agriculture, from the difficulty of waste disposal, and most obviously from the lack of clean water. In addition, there is a lot of dust in the air, partly because the village streets and paths are unpaved.

Discovery Procedures

First Survey

We carried out the first survey in spring 1995 after some delay in obtaining the research permit. The research areas were chosen for their natural boundaries or some administrative separation. Within each broader area, we chose a street or block to include roughly the number of households that we wanted. The interviewers were instructed to choose one adult from within each household. Furthermore, they were to seek a balance of males and females. There was often not much choice since only one adult was at home at the time of the team's visit, so the team adjusted the sample composition by varying the times of their visits. In Dar al-Salam and Sayyida Zeinab this produced a roughly equal gender division. The overall sample size was 2,266 persons.

The samples are not representative. We chose the research sites because they did not have any outstanding characteristic that would make them atypical, and then tried to interview each household in a street or group of streets. We deliberately made the sample relatively large so that the overall size of the sample would to some extent compensate for the lack of representivity.

Resident Researchers

Our plan was to have one team member responsible for each of the sites. If possible, this team member was to live in or near the site, and to be our expert on that site. The resident researcher's tasks included observing participants, making supplementary interviews, noting institutions and activities in the areas, organizing the focus groups, and helping the other research teams in their activities. The resident researchers periodically delivered their field notes and other documentation, and some of this material was then translated from Arabic into English. We held regular biweekly meetings during which we discussed the progress of the research.

In practice the only way to find someone to live in the areas was to recruit from the areas. Local men served as resident researchers in the three urban areas and a young woman in Minufiyya. All were university graduates.

Focus Groups

Our plan was to organize at least four focus groups, the population divided into men and women, older and younger, in each of the four areas. We also organized other focus groups: one of workers in the cement company in Kafr al-Elow, one of workers in a battery plant in Dar al-Salam that had recently closed down, one of university degree holders who had moved away from Sayyida Zeinab to see how they would contrast their new home with their old, one of farmers distinct from senior men in Abkhas, and also one of young girls living in an *'ezba* instead of in the main village. Altogether we carried out 22 focus groups in the four research areas between November 1995 and January 1996, and in June and November 1996 two focus groups of women in Ma'adi. The discussions generally lasted about an hour and were taped. The tapes were transcribed in Arabic, and English summaries were prepared.

Measuring Pollution

If we wanted to understand people's opinions, we had to have solid information on the environmental conditions to which they were reacting. A team of engineers joined the project to identify and measure sources of pollution in our research areas. With the help of several colleagues, they carried out an environmental audit in the four localities and then focused on measuring air and water quality and noise levels.

Second Survey

When we carried out the second survey in 1997, we returned to the same areas and followed the same techniques. The original plan was to carry out a second survey with a longer questionnaire administered to fewer people. However, in the end we retained the same sample size and instead avoided some of the complicated questions asked the first time while probing in particular some additional points we had become aware of during the first year of the study. The second survey was carried out in early 1997, about two years after the first one. The sample size was 2,307 subjects.

Description of the Samples

In the first survey we interviewed 2,266 individuals, each one from a different household. The three Cairo localities had about the same number of respondents, while the village of Abkhas had fewer, reflecting a near total sample of households. Overall, the gender responses were about equally divided. Table 1 gives the breakdown of the two samples by locality, gender, age, and education, using the categories that we return to in future analysis. We carried out a more detailed analysis of the first sample, in 1995, and describe that below. For the 1997 sample we show that it contains comparable proportions of our main categories, but did not collect again detailed information on the identities of the respondents. We argue that the two sam-

Table 1: Categories of respondent samples

	1995		*1997*	
	n	%	*n*	%
Locality				
KE	601	26.5	601	26.1
DS	594	26.2	601	26.1
SZ	601	26.5	599	26.0
AB	470	20.7	506	21.9
Gender				
Female	1,107	48.9	1,129	48.9
Male	1,159	51.1	1,178	51.1
Age				
≤25	432	19.1	511	22.1
26–35	659	29.1	629	27.3
36–45	535	23.6	496	21.5
≥46	640	28.2	671	29.1
Educational status				
Without formal schooling				
Illiterate	615	27.1	614	26.6
Can read and write	378	16.7	285	12.4
Schooling completed				
Primary	174	7.7	196	8.5
Preparatory	193	8.5	213	9.2
Secondary	549	24.2	515	27.3
Postsecondary	357	15.8	484	21.0
Total	2,266		2,307	

KE=Kafr al-Elow, DS=Dar al-Salam, SZ=Sayyida Zeinab, AB=Abkhas.

ples are essentially identical in terms of their social composition, and so we have combined the results in a single presentation while being careful to distinguish the two surveys.

Table 1 gives the breakdown of the major categories in both the 1995 sample and the 1997 sample.

It is interesting to compare these categories with each other. Of the four localities, Dar al-Salam and Sayyida Zeinab had almost the same gender breakdown as the overall sample (male:female ratio of 51:49), while Kafr al-Elow had more females (45:55, male:female) and Abkhas more males (58:42, male:female). If we arbitrarily use the midpoint in our age scale (≥35; ≤36) as a guide to the age-related make up of each community as a whole, then Sayyida Zeinab had the oldest population (roughly 37%<36), followed by Dar al-Salam (48%<36). Abkhas (51%<36) and Kafr al-Elow (58%<36) had more people younger than this arbitrary cutoff age. The median age was about 36 years. In terms of education, Sayyida Zeinab has the highest proportion with secondary and university education, and the fewest illiterates, while Kafr al-Elow has the greatest number of illiterates and the largest number who can only read and write. If, for our sample, we take the proportion of "unschooled" (illiterates and those who can read and write but without schooling), the ratio is 56.2% in Kafr al-Elow, 52.3% in Abkhas, 38.9% in Dar al-Salam, and 29.8% in Sayyida Zeinab. Kafr al-Elow has the youngest, the most female, and the least schooled respondents of the four subsamples.

The Kafr al-Elow sample has the largest proportion of females, who are less likely to be schooled. In fact, the unschooled make up 35.2% of the male population and 52.9% of the female, including the 16.3% of the males and the 38.5% of the females who are illiterate. In other words, those who can "only" read and write outnumber the illiterates among the males but not among the females. The proportion of females among the illiterate ranges from 62% in Abkhas to 76.4% in Sayyida Zeinab, with Dar al-Salam at 65% and Kafr al-Elow at 73.5%. Thus, about two thirds of the illiterate are female.

The older groups in our sample tend to have a preponderance of males. The youngest age group is 40% male, and the oldest 59.5% male, with the others in between. The older groups also tend to be less educated: the unschooled are 18.2% of the youngest group, 37.5% of the second, 45.2% of the third, and 66.6% of the oldest group.

Among our respondents, 870 women identified themselves as housewives (38.4% of the total sample and 78.6% of all women). Another 206 individuals (9.1%) were on pension (20.9% of them women), and 152 (6.7%) were students (44.7% of them women). So just over half the sample (54.2%) were housewives, retired, or students. Of those in the workforce, we classified 560 (24.7%) as working class and 477 (21.1%) as middle class, on the basis of job titles and the probable level of education required for that job. Among the workers there were farmers, laborers, mechanics, drivers, carpenters, painters, tailors, barbers, bakers, grocers, guards, and also those who were serving in the army and the unemployed (17 males and 15 females). Among the middle class there were white collar workers, employees, teachers, shop owners, army officers, accountants, doctors, nurses, lab technicians, and lawyers. The highest in the social scale were doubtless the eight doctors. The sample could thus be said to represent largely the working and the middle class.

Factory workers predominate in Kafr al-Elow: they are nearly half of the employed. The two largest employers of residents of Kafr al-Elow are the Helwan Portland Cement Co. and the Iron and Steel Co. There are still a few farmers. The remaining workers include merchants, professionals such as doctors and engineers, bureaucrats, and "quasi-proletarians." The young people have a problem getting jobs, because there are few new openings in the nearby factories, whether public or private. Some were hired on a temporary basis by the nearby factories, and hoped that this would lead to a permanent job, but it rarely did. Temporary workers work full time but with lower pay and less security, and they often feel quite badly treated. One temporary worker in the Delta Steel Co. noted, "I have a family and could not take the risk of working one day and not working for 10 days. So I decided to work in the factory when I was offered the job. It is a humiliating job. We are treated like dogs. The temporary workers are really crushed. We do not have any rights." Other young men have had short-term jobs in construction or as unskilled labor. Some young people have diplomas or even university degrees, and they wait for a different kind of job or work with their fathers in a family business.

We asked people how long they had lived in their neighborhood, and naturally some were born there. We also asked people why they lived where they did, and again some answered that they were born there. The second figure, reflecting choice, was slightly lower than the first, the biggest spread (5 percentage points) being in Kafr al-Elow. Overall, 50% of the respondents said they were born in the neighborhood where they lived (though we did not challenge them to define that neighborhood), but the range was from 93% in Abkhas village to 10.9% in the relatively new area of Dar al-Salam, with Sayyida Zeinab (56.7%) and Kafr al-Elow (48.4%) falling in between. Kafr al-Elow had the largest number of people who said they lived there or had moved there because it was near their work (21.6%), followed by Dar al-

Salam (7.6%) and Sayyida Zeinab (2.7%). No one in Abkhas gave this answer.

The population is generally stable in terms of residence and jobs. While half the respondents said they were born in the locality where they were interviewed, another 36.2% said they had lived there at least 10 years. Only 5.4% had moved in within the previous three years. Of the half born "elsewhere," most (601 of 1,132, or 53.1%) were born in Cairo. The others represented a wide range of Egypt's governorates, covering everywhere except Sinai and Marsa Matruh. Those people who had moved most commonly gave "marriage" as the reason, followed by "work." Only relatively few seem to have made a decision to move based on the attributes of the neighborhood and its potential.

Only a quarter (25.4%) said that their work was in a different place than their residence. Of these the respondents in Dar al-Salam and Sayyida Zeinab were most likely to work in one place and live in another (35.9% and 31.8% respectively), compared to Kafr al-Elow and Abkhas (19.3% and 11.9%). Of the 1,034 individuals with jobs, 980 had not changed jobs in the previous five years; the 54 who had were mostly in search of a better (or better-paying) job.

Belonging to a formal organization was uncommon: 91% of the respondents said they did not belong to any organization at all, while 6.3% (n=143) said they belonged to a sporting club and 2.1% (n=47) to a labor union. Another 11 individuals belonged to other organizations, such as a development society. We did not record any overlap in these memberships. Membership was most common in Sayyida Zeinab (16.3% of the sample), followed by Dar al-Salam (10.6%), Kafr al-Elow (5.8%), and Abkhas (1.7%). Very few people mentioned any leisure-time activities; mostly this was conceived as sitting at home with relatives and watching television. Only in Abkhas village did people identify more active leisure activities such as fishing, playing ball, or going to the fields.

The media exposure of the sample is important as an indicator of the integration of individuals into the Egyptian media economy, which can be taken as a quality of life indicator, and also because the media are a source of information and attitudes about the environment. Table 2 gives the basic breakdown on exposure to newspapers, magazines, and television for our sample.

Most of the 1995 respondents (91.3%) said they watched TV, primarily soap operas and Arabic films. More than half of these (49.7% of the subsample) said they watched more than two hours a day, while 11.2% said they watched more than five hours a day.

Nearly half (1,020, or 45%) of the respondents said they read a newspaper.[23] Among these, the most popular newspaper was *al-Akhbar*, mentioned by 70.8% of the 1,020 individuals. Following it was *al-Ahram* with 45.7%. The others mentioned, but falling off rapidly in numbers were *al-Gumhuriya, al-Wafd, al-Sha'ab, al-Ahrar*, and *al-Ahaly*, plus some others such *al-Masa'*, the evening *al-Ahram*, and *al-'Alam al-Yaoum*, the newspaper of the armed forces. The proportion of those who read a newspaper varies a great deal between the different localities. Thus, the highest rate is in Sayyida Zeinab (67.3%), followed by Dar al-Salam (54.7%). Kafr al-Elow (35.8%) and Abkhas (20.6%) are lower.

Table 2: Percentage of respondents exposed to various media (1995)

Medium	%
Newspaper	45.0
Magazine	23.7
Television	91.3
≤2 hrs/d	41.5
>2 hrs/d	49.7

Source: 1995 survey. Percentages are of the whole sample (N_{1995}=2,266).

Fewer people read magazines than newspapers. Only 23.7% (526) of the respondents mentioned a magazine. The most popular magazine was *Nusf al-Dunya* (123, or 23.4% of those responding), followed by *Akher Sa'a* (76, or 14.4%), *al-Musawwar* (42, or 8%), *Oktober* (also 42, or 8%), and *Kalam al-Nas* (6, or 1.1%).

There is a striking gender difference in exposure to the media, although men and women are equally likely to cite the media in general as their source of information in learning about the environment. While a slight majority of men (55.4%) say they read a newspaper, only about one third of women (34.2%) say they do. Conversely, 59.3% of women say they watch more than two hours of television a day, compared with 40.7% of the men. In sum, men are more likely to read a newspaper, and women to watch television, as a source of information. This is consistent with differences in literacy rates.

Altogether 156 individuals (6.9% of the total sample) said they neither read newspapers nor watched television. The largest group of these is in Abkhas, where they were 13.8% of the local sample. At the other extreme, 543 (24%) said they both read newspapers and watched three or more hours of television every day. This group was most numerous in Dar al-Salam and in Sayyida Zeinab (33.1% and 32.5% of the local sample, respectively).

There is a slight tendency for television viewership to rise with newspaper reading. Those who read newspapers are 21.2% of the nonwatchers, 46.2% of the low watchers, and 48.2% of the high watchers.

Data Analysis

Each of the various streams of data implied its own form of analysis, and these have been combined in this book into a single argument. The general thread of the argument is to start from a descriptive analysis of the environmental conditions under which people live. We move from there to the cultural construction of the environment and the theme of major concern, environmental pollution. Finally we pass on to the analysis of the different forms of action, individual and collective, before concluding. In short, we move from an understanding of conditions to the level of ideas before reaching an analysis of political and social action.

The results from the two surveys have been analyzed not only as results from the sample as a whole, but according to various cross-cutting categories. The one that consistently proved the most interesting was the breakdown by locality, and many of the data are presented in this form. There was sometimes but not always a contrast between the survey results according to gender; we have presented cases both where there was and where there was not a contrast, since gender is a major axis of difference in Egyptian society, as elsewhere. The other variables (education, age, and occupation) did not consistently produce additional enlightenment, and are only occasionally invoked. In presenting the data, we have used percentages for the most part to facilitate comparison between the different results.

Generally we used contrastive analysis in the survey material to avoid quick answers and stereotypes. Thus, we often asked our respondents to compare different situations rather than asking directly for an evaluation of those situations, and we compared the results of different questions with each other. In our presentation of data from the surveys, we usually break the figures down by locality, and sometimes by gender or other independent variables, thus introducing a comparative dimension at that level. When there are relatively few differences between localities, genders, or other factors, we take it that we are dealing with a cultural model. Where there are specificities in the data from a particular locality, this can generally be attributed to the different circumstances in each place.

The data from the focus groups, interviews, and observation are incorporated with the survey data. In some areas, such as an understanding of living conditions and the political action that people may take, these data are predominant. Such data are essentially narrative in form, and thus contrast with the more structural outcome of the survey data. Our data support the argument that views on environment and pollution are cultural constructions, and that action occurs as an outcome of these cultural constructions. It was one of our goals in this project to amalgamate all these streams of data into this study.

3

Measuring Pollution

The first problems are garbage and sewage and pesticide pollution, because they affect all Egyptians. We in the country and those in the city both have these problems. Then water and air pollution, because they affect some of us. As for nuclear pollution and ozone layer destruction, these are not our problems. They belong to other places of the world. We have no such technological development to cause such problems.

Two male students from Tanta, studying in Cairo, 1993

In presenting the results of the measurements of air and water quality and of noise levels in the four research localities, our goal is to show the similarities and differences between the research localities in terms of actual quality of the environmental conditions. This information provides background for the attitudes, opinions, and experiences described by our respondents in these localities. This is the "real" situation to which they were reacting. While the data assembled by our research team may differ in detail from data collected elsewhere in greater Cairo, it is consistent with it (see Project in Development and the Environment (PRIDE) 1994: vol I, pp. III:6–40; MSEA 2000). Generally speaking, our data show that the most affected environment is in Dar al-Salam, followed by Kafr al-Elow, although water is the most problematic in Abkhas.

Measurements

Solid and Liquid Waste

We could not measure what people consider the worst problem (garbage and sewage in the streets). Some overall estimates for solid waste in Cairo are available, generally in the range of 10,000 tons per day (PRIDE 1994, vol. I, p. III:17) or more.[24] The amount increases with population and development, and it is also said that the amount of solid waste per person in Cairo has doubled in the past 25 years. About two thirds of the solid waste is from the residential sector, and most of the rest is evenly divided between the commercial and municipal sectors (mostly street cleaning). A bit less than half the solid waste is food (46%). Paper is another large component at 21%, and the remainder includes metals, plastic, and other materials. About half the solid waste is collected and managed, with much higher rates in upper-income areas where there are more recyclables. The uncollected waste is either dumped in canals or on land or burned on site (ibid.). In the early 1990s, Cairo had no operating sanitary landfills, so what could not be recycled was eventually dumped in open dumps in the desert. Collection and disposal of garbage have been improved somewhat since the early 1990s, although it remains, if one may say, a contested terrain (Volpi 1997).

As for sewage, Cairo was estimated in the early 1990s to generate 2–3 million cubic meters (mcm) of domestic sewage and 0.2 mcm of industrial effluents daily (el Gohary 1994). Most of this is discharged into canals and drains toward the Mediterranean. Most industry in Helwan is not allowed to discharge water to water-treatment plants without pretreatment, so the treatment plants are underutilized (PRIDE vol. II, p. B-2, B-7). El-Gohary also notes, "with the exception of fecal coliform bacteria, the levels of contaminants in both Nile water and agricultural drains are well within guidelines for safe irrigation of food crops. It appears that sources of irrigation water in greater Cairo may exceed desirable levels of bacteriological contamination" (El-Gohary 1994: B-19; see also El-Gohary, Nasr, and El-Hawaary 1998).

Sampling the Air

We collected air samples from the four localities in our study. The samples were collected using a high-volume air sampler from the different sites, different locations on each site, and at different times during 1996–1997. They were analyzed for total suspended particulates (TSP), lead, smoke, sulfates, chlorides, nitrates, and free silica.

In line with other findings, we assumed that industry and increasing traffic were a substantial source of air pollution in Cairo. The problem of industrial air pollution was worsening due to the continuing lack of attention given to the siting of industrial plants, or the lack of provision and operation of air pollution–control equipment. On the other hand, our measures were taken before the lead content of gasoline in Egypt was reduced in favor of MTBE (see O'Toole et al. 1996). At the time of our research, public information on air pollution was not generally available. The "black cloud" episode of fall 1999 shows that access to information continues to be a problem. Nor was an inventory of industrial emissions possible when more than 85% of them were uncontrolled because of lack of regulations and financial commitment.

A main source of air pollutants is emissions from vehicles, among them at the time of our study leaded gasoline. Atmospheric lead, from fuel and other sources, is harmful to humans, especially children and the elderly.[25] Unburned and waste hydrocarbons (HC-s) from fuel combustion combine with nitrogen oxides in the presence of sunlight and form a complex variety of secondary pollutants called photochemical oxidants, or smog. Nitrogen dioxide (NO_2), for instance, is a poisonous brown gas, whereas ozone (O_3) is highly reactive and can damage the lungs. Burning high-sulfur diesel fuel forms sulfur oxides (SO_x) that cause temporary and permanent injury to the respiratory system. Particulates are another major pollutant that is common in Egypt due to the unpaved roads, the closeness of industrial sectors to the residential communities, and the inefficient use of fuel in transport. Particulates are absorbed into the body through the respiratory system, and can have severe effects, especially on smokers and others with respiratory conditions.

Table 3: Ranges of TSP, smoke, and lead concentrations

Locality	*TSP* $\mu g/m^3$	*Smoke* $\mu g/m^3$	*Lead* $\mu g/m^3$
KE	450–650	67–144	0.06–1.15
DS	551–1,012	86–150	0.27–1.02
SZ	412–586	66–120	0.18–1.26
AB	289–470	39–63	0.01–0.11
WHO standard	75/year	60/year	0.4/year
Egyptian standard	90/year	60/year	1.0/year

KE=Kafr al-Elow, DS=Dar al-Salam, SZ=Sayyida Zeinab, AB=Abkhas.

Source: 1995 team measurements.

The range is from the highest to the lowest of our measurements. Almost all measurements in the research localities were above WHO and Egyptian standards.

Our research concentrated on three major types of pollutants: TSP, smoke, and lead (see Table 3). Contrary to the usual assumption in Cairo, our evidence shows that the air is generally more polluted in Dar al-Salam than in Kafr al-Elow, near Helwan. Dar al-Salam appears as the most polluted on all three main measures (TSP, smoke, lead). Sayyida Zeinab is about equivalent to Kafr al-Elow (TSP level is lower; smoke and lead levels are higher), while Abkhas generally has the least polluted air.[26]

The recorded value for lead tended to be above WHO benchmarks and the Egyptian standards in the three urban areas but was acceptable in the village. However, the recorded site values of TSP levels were much higher than the standards (4.4 to 12 times higher). Smoke is within acceptable limits in the rural site, but is above the international standard in the urban and semi-urban communities. Other chemicals we measured in airborne dust were within international norms: sulfate, nitrate, free silicate, and chloride.

Water Quality

Water quality also differs from one locality to another. We collected samples from house taps in Kafr al-Elow, Dar al-Salam, and Sayyida Zeinab, and from a public tap in Abkhas. Despite poor conditions around the public tap, the

Table 4: Pollutant concentrations in water from various localities

	Pollutant		
Sample site	*Coliform (cell/100 ml)*	*Soluble salts (mg/l)*	*Hardness ($CaCO_3$ mg/l)*
KE public tap	0	298	not measured
KE house tap	0	301	not measured
DS	0	282–288	not measured
SZ	0	288–294	not measured
KE canal	$2.7–9.0x10^3$	—	—
Subk treatment plant	30	620–626	390–395
AB hand pumps	0–600	368–436	180–275
AB canal	70	270	170

KE=Kafr al-Elow, DS=Dar al-Salam, SZ=Sayyida Zeinab, AB=Abkhas.

Source: 1995 team measurements.

water was comparable to that in the house taps. In Table 4 we see that the samples are free from bacterial contamination and safe bacteriologically to drink and use for other household purposes. Ammonia and nitrates are below the acceptable standard maximum values indicating that the water sampled is free from extraneous sources of pollution such as possible contamination in the roof tanks.

The Khashab Canal in Kafr al-Elow

The Khashab Canal runs through Kafr al-Elow. Its water is used to irrigate the agricultural land in the surrounding area, and the canal is used as a dump for household and industrial waste. It is essentially an urban canal. Water pollution in the canal was assessed through collection of water samples from five sites along the canal representing the different conditions prevailing in the research area.

The bacterial contents of water samples indicate human and animal wastes mixed with the canal water. Discharged raw sewage contributes to the pollution of the canal water as is evident from the presence of pathogenic bacteria, such as coliform, salmonella, and shigella, in the water samples. In five samples, coliform bacteria ranged from 2.7×10^3 cells/100ml to 9×10^3 cells/100ml, and salmonella and shigella ranged from 1.2×10^2 to 2.8×10^2 cells/100ml. Salmonella and shigella also confirm the presence of raw sewage as a source of pollution, as they are human pathogens that induce gastrointestinal health problems in humans. The impact of these pathogens could be fatal if people, especially children and the elderly, come into contact with infected water—for instance, if children wash or wade in the canal on hot summer days.

The water from the Khashab canal is highly contaminated and it constitutes a health hazard even for people coming into contact with such water through irrigation. Polluted water is indicated by the dissolved oxygen tension and the relatively high values of biochemical and chemical oxygen demands (BOD and COD) created by dumping solid wastes and domestic raw wastes generated in the nearby residential housing areas. It has some biological toxicity as the observed absence of phytoplanktons or aquatic plants in the sector facing the residential area confirms, yet the water quality is within norms for use in agriculture.

Although canal water is not suitable for human use, it is within the range for irrigation water (see also El Gohary et al. 1998), based on measurements of sodium adsorption ratio, the pH values (which are around neutrality), and the relative absence of harmful heavy metals such as manganese, copper, lead, and nickel. Collected water samples can be classified as a high salinity hazard and low alkalinity hazard (C_3-S_1 grade), while the Nile water is classified as a medium salinity hazard and low alkalinity hazard (C_2-S_1 grade).

Water in Abkhas

All sources of water in Abkhas are problematic. This finding makes Abkhas similar to other rural areas in the Delta region (see El Katsha et al. 1989). In Abkhas, there are three sources of water: (1) the government system of treated water, piped in from tanks in the mother village of Subk al-Dhahak; (2) pumps, reaching ground water at about 20m–30m; and (3) canal water, mostly from the large Bahr Shebin, one of the main irrigation canals in the Delta, and some from smaller irrigation canals whose offtake is outside the village. We took samples from the pumps and the canal, from the house taps and from the treatment plant in the nearby village of Subk.

The government-operated water distribution system, which began in 1989, provides drinking water from a water-treatment plant serving several villages. The managers of the government water system in Subk al-Dhahak stated that the well drew from 60m–65m below the surface, that they used 1,800 kg of chlorine per month, and that the tank was washed out with detergent at least once a month. They described complaints as relating mostly to the bad smell of the water. "The percentage of manganese and maybe iron in the water is high. We keep reporting that and nothing happens." They also pointed out the problem whenever drainage pipes break open near water pipes and the waters mix. Another problem is seepage from tanks beneath resident's houses. In 1994 the local council replaced 4 km of drinking water pipes at a cost of LE 176,000, drawn from the Social Fund.

When we were doing our research, the piped water came from a deep well in Subk al-Dhahak, and was pumped from the well to a reservoir from which the water flowed through the pipes to taps in Subk al-Dhahak, Abkhas, and the other connected villages. Toward the end of our research project, the government was building a pumping and purification station on the Bahr Shebin itself that would supply water to the same reservoir and pipe network as the well. The piped water network does not extend to the hamlets on the far side of the Bahr Shebin (Ezbat al-Bahr and Tajammu' al-Kubri). The inhabitants of these hamlets depend on water pumps and the canal for water. Some inhabitants even draw their drinking water from the canal.

The characteristics of the treated water as it emerges from the plant are not satisfactory; total bacterial counts are high, reaching $7x10^5$ cells/ml, and the *Escherichia coli* bacterial count is 30 cells/100ml. Bacteria should be completely absent in drinking water to satisfy national drinking water standards. The high *E.Coli* count indicates that the intake water is highly contaminated with domestic and animal wastes and also that the disinfection process in the treatment plant is not efficient. Moreover, the data indicate that by the time the water reaches the consumer in the village, its bacterial quality has further deteriorated as a result of passing through a poorly maintained distribution system. This finding is also supported by the high suspended-solid content in the sampled water, which increased from 44 mg/l in the water as delivered from the plant to 134 mg/l in the water sample collected at the house tap. Recorded ammonia concentrations of 0.9 mg/l also support the above finding as the presence of ammonia indicates nearby sources of contamination.

The pumped underground water was found to have higher levels of soluble salts (TDS) than those found in the canal water. These levels are additionally high since domestic wastewater mixes with the aquifer water, leading to contamination of the pumped water, and bringing a threat of higher morbidity. The water from the treatment plant is the "hardest," measuring 390–395 ($CaCO_3$, mg/l), while the water from the pumps varies from 180 to 275 mg/l, and the canal water is the "softest," at 170 mg/l.

Noise

We measured noise levels in several parts of the study localities. Noise was recorded as an average of at least 15 measurements taken by a portable instrument while standing still at different sites. Standards for permissible noise exposure vary according to length of exposure. For instance, exposure to 85 dB for more than 8 hours a day produces permanent damage (Miller 1998: 320). Table 5 shows our measurements.

Generally our results show that noise levels are close to the permissible limits, but there is considerable variation within all localities as a function of ongoing activities, and some areas, notably Dar al-Salam, exceed the limits.

Table 5: Average noise levels at the study locations (Daily ranges)

Locality	*Noise Level (dB)*
KE	50–85
DS	60–95
SZ	60–90
AB	50–85

KE=Kafr al-Elow, DS=Dar al-Salam, SZ=Sayyida Zeinab, AB=Abkhas.

Source: 1995 team measurements.

Overall, Abkhas is the "quietest" locality, followed by Kafr al-Elow, where the noise level recorded in the quiet residential zone was 50 dB. The noisiest locality is Dar al-Salam, particularly near the Metro station (al-Mal'a) where there are many shops and crowds (95 dB). A side street in al-Gezira experienced the lowest noise levels in Dar al-Salam with recorded readings ranging from 60 dB to 65 dB, still more than Kafr al-Elow or Abkhas. However, all localities can sometimes be dangerously noisy, depending on the activities occurring.

Chemicals in Agriculture

Chemical fertilizers and insecticides are a major environmental problem in rural Egypt. The USAID report notes that pesticide intake through food is apparently two orders of magnitude higher than in other countries, perhaps because of remaining residues from discontinued pesticides such as DDT, aldrin, and dieldrin, all three banned in Egypt in the early 1980s. "The pesticides that have been used in Egypt for the past decade (organophosphorus, carbamates, and pyrethroids) are much less long-lasting than organochlorines and most are not thought to be carcinogenic . . . . They may pose fewer risks than residues of 20-year-old organochlorines" (PRIDE vol. 2, E:5–6; see also Euroconsult 1992:67). DDT derivatives can last up to 35 years, compared with 18 months for urea, and about three months for carbamates and organophosphorus, while pyrethroids are rapidly degraded by the sun on the soil surface.

Comparison

The results of our study are consistent with other analyses of the same phenomena. The overall status of the four localities is fair to poor. Dar al-Salam has the worst record for air and noise pollution, and in both cases is worse than international norms, but is acceptable with respect to drinking water purity. Abkhas has the worst situation with regard to polluted water, since none of the three main sources of water is fit to drink, and the air has a high TSP level, albeit the best of the four sites; however, noise levels are acceptably low. Kafr al-Elow has a serious problem with air and water, but somewhat less with regard to noise, though the area along the Corniche highway is noisier than the traditional settlement. Sayyida Zeinab has the second highest noise level, has a serious air pollution problem, but is acceptable with regard to drinking water. These estimates are summed up in Table 6. It should be kept in mind that these are relative rankings rather than absolute values.

Unfortunately, we could not develop a numeric index for garbage and sewage in the streets, which was one of the main complaints people had. It is thus hard to rank the four areas according to the prevalence of solid and liquid waste.

Table 6: Comparative environmental fitness ranking of the research localities

Locality	*Air*	*Water*	*Noise*	*Total*
DS	1	3	1	5
KE	2	2	3	7
SZ	3	4	2	9
AB	4	1	4	9

1 = most polluted. The lower the ranking, the more polluted.

The total score is the sum of the numbers in the first three columns.

KE=Kafr al-Elow, DS=Dar al-Salam, SZ=Sayyida Zeinab, AB=Abkhas.

4

Environmental Conditions in the Research Localities

The part of Sayyida Zeinab I live in has little or no air pollution. We have no nearby factory. We have good water. It sometimes is cut but this is not common. Moreover we have no garbage problem. The street we live in is not really a street, it is an alley, and it is clean. The shop owners clean every day, they make sure that there is no dust or dirt or garbage. We houseowners give our garbage to the garbage collector who comes regularly every day, and if he does not come, we argue with him. Do you know that every father in this very small street beats his son if anybody reports that he has thrown garbage in the street. This is exceptional. We manage to do this because we are few and because we know each other and because the shop owners are keen to have the street clean. They are about three in number. But because the street is short and narrow they are able to keep it clean . . .

Yet if you go to other places or even to the main square you find garbage and dust in the street. Therefore, if I say we have few problems, that does not necessarily apply to other places in Sayyida Zeinab. We have few problems because we are managing to solve our problems.

Taxi driver, Sayyida Zeinab, 1993

In this chapter we provide a descriptive analysis of the living conditions in our four research localities, based on answers to our survey questions, insights from the focus groups and the interviews, and observations. Our presentation covers housing conditions, neighborhood life, the disposal of solid and liquid domestic waste, the state of water, air, and noise pollution, and such other problems as pests (insects and rodents), chemicals, and moral pollution. In addition to the physical facts of the case, our data also reflect the attitudes of our respondents and participants. The picture for Cairo and our rural site is broadly consistent with accounts of other major third world cities and countrysides.

Standard of Living

Housing

Most housing in Cairo is informal, usually in the sense that it is built without a formal permit, occasionally in the sense that title to land is not clear (Steinberg 1990, Ismail 1996, Al-Hadini 1999; El-Kadi 1987; Ireton 1988). There is often a delay in the provision of utilities such as water, electricity, sewage, or telephone service, and so urban areas vary according to the level of amenities.

Table 7: Amenities by locality

	Locality				
	KE	*DS*	*SZ*	*AB*	*All*
Amenity			*(%)*		
Piped water	96.7	99.5	100.0	43.9	86.7
Drinking water from govt. system	95.7	95.3	96.5	1.0	75.0
Sewage connection	58.2	99.8	99.5	0.4	67.1
Paid for connection*	82.9	44.2	3.2	—	37.1
Pay for maintenance**	27.7	10.5	4.7	—	12.1
Cesspit access†	33.8	0.1	—	66.1	97.1
Regular garbage collection	3.7	98.0	99.2	0.4	52.3
Feelings about neighborhood					
Dustier than formerly	98.3	55.1	55.3	70.6	69.8
Noisy	36.3	54.4	59.1	4.7	40.0

* & ** n=1,548

† n =759. Figures for cesspit access are percentages of those without sewage connections.

KE=Kafr al-Elow, DS=Dar al-Salam, SZ=Sayyida Zeinab, AB=Abkhas.

Source: 1997 survey. Percentages are of the whole sample (N_{1997}=2,307) except where indicated.

A summary of the data on amenities for 1997 is given in Table 7, and discussed below.

Although they are broadly consistent with other reports from contemporary Cairo, the data are a description of the circumstances reported by our sample, and should not be generalized to any wider group. Some of the data refer to the 1995 survey, and some to the 1997 survey, as indicated.[27]

Overall Sayyida Zeinab appears as the best equipped and richest neighborhood, followed by Dar al-Salam. Kafr al-Elow is the poorest of the three urban localities, and the least well equipped. Abkhas households have pretty much the same range of equipment as the urban ones, including electricity and toilets, a little more space on average, but are less well integrated into the water and sewage systems. Abkhas dwellings are on average larger and newer than the urban ones. All households were connected to the electrical system in Sayyida Zeinab and Dar al-Salam, and nearly all in Kafr al-Elow (98.3%) and Abkhas (97.7%).

Almost all households were equipped with washing machine (94%), television (93.8%), radio (91.5%), *butagaz* (bottled gas) (90.2%), and refrigerator (85.6%). Each of these five items was present in at least 80% of households in each of the three urban localities, while Abkhas had similar levels for the first three but lower levels for refrigerators (65.9%) and *butagaz* (77%). Constructing a score for these five items, the order from best equipped to least well equipped is Sayyida Zeinab, Dar al-Salam, Kafr al-Elow, Abkhas.

Telephones are less widespread (27.5% overall). The telephone was present in 65.5% of Sayyida Zeinab households in our sample, reflecting an older and more middle-class neighborhood, but only 20.7% of Dar al-Salam households, 11.3% of Kafr al-Elow households, and 8.1% of Abkhas households.[28]

Overall, 81.6% of the respondents lived in houses that were more than 15 years old (i.e., built in 1980 or earlier). The locality with the fewest old houses was Abkhas (68.5%), that with the most was Sayyida Zeinab (93.8%), with Dar al-Salam (83.2%) and Kafr al-Elow (77.9%) in between. The average number of rooms per household was 3.46. Dar al-Salam (3.39) and Sayyida Zeinab (3.43) were very close to the overall average, while Abkhas (3.94) was above average and Kafr al-Elow (3.17) below average.

Another insight into the respondents' sense of their standard of living can be gained from the improvements in their surroundings they thought were necessary. People were most likely to think that the streets needed paving or cleaning, and that they needed to be hooked up to the sewage system. Relatively few wanted improvements with regard to water (cleaner water), schools (more of them), transportation (more buses), or electricity (fewer cuts).

Driving or having access to a motor vehicle is another indicator of standard of living. In our 1997 sample 236 people (10.2%) said they drove a car or a motorcycle. They were most numerous in Sayyida Zeinab (15.4%) and least in Kafr al-Elow (5.5%), with Dar al-Salam at 11% and Abkhas at 8.9%. The drivers were 13.8% of the men and 6.6% of the women. Most of those who said they drove a vehicle also owned it (75%).

Feelings about the Neighborhood

Most people surveyed said that they like the neighborhood where they live because the presence of relatives (37.2%) and friends (29.3%) made it pleasant.[29] Men were more likely to note that their home was near their work (16.2% of men, 8.1% of women), while women were more likely to say that there was nothing about the neighborhood that they liked (20.3% to 12.5%). Less than 10% gave an economic reason for liking their neighborhood (e.g., the neighborhood was good because it was inexpensive). Those with higher educational levels were somewhat more likely to cite the presence of friends, and those with lower educational status somewhat more likely to cite the presence of relatives.[30]

People in the focus groups were generally positive about their neighborhoods, frequently citing either convenience or their personal networks of family and friends as the reason. Others were more ambivalent, and cited such negative aspects as growing pollution or social deterioration.

In Kafr al-Elow, some people are attached to the village/neighborhood as the home of their families but others would leave if they could afford to. On the positive side, there is community spirit and neighbors are friendly. Visiting is common. High-quality fruits and vegetables are available in the market, and outsiders come to shop. (Though others complain that the wares in the market are covered with cement dust.) On the negative side, focus group members referred to the time when Helwan was a resort area, and Kafr al-Elow was an agricultural village. Now Kafr al-Elow has become an industrial area and a slum, and more heterogeneous socially. A further aspect of this decline is noise along the Corniche where workshops and truck stops are concentrated.

People living in Dar al-Salam say they like the easy access to transport 24 hours a day, and the good shopping with all commodities available. Some Dar al-Salam residents professed to like the crowds of people, since it gave them a sense of security, but others were less comfortable with the population density (which gives Dar al-Salam its nickname of "People's China"). Many people cited "friendly neighbors" as a reason to like Dar al-Salam, though others thought they were inconsiderate, and thought that the growth of the area has led to moral degeneration. Only some among the younger people wanted to move away. When people compare themselves with Ma'adi or Zamalek they feel underprivileged and neglected.

In general the people of Sayyida Zeinab appreciate the central location of their neighborhood, the availability of all goods and services, the ease of transportation, and the amicable relations and a sense of care and belonging among the residents. It is a "popular" area in the meaning analysed by Diane Singerman (1997:11), that is,"sha'bi," associated with the people or folk, friendly and familiar. There is a degree of cooperation that may be lacking elsewhere. There is a feeling of security because other people are constantly present in the streets. People are conscious of the history of the area, and appreciate the religious atmosphere conveyed by the large number of mosques.

In Abkhas people stressed that the air is cleaner than in cities because of the absence of factories, buses, and cars, though dust is a problem. The young men in particular feel strongly about the absence of paved roads in the village, which makes walking dusty. The *'ezbas* are generally cleaner than the village because the population is lower and the fields are closer. Yet by the same token, the *'ezba* residents are more exposed to the negative effects of the use of fertilizers and chemicals in agriculture.

Participants in a focus group of those who had moved from Sayyida Zeinab to Sixth of October City (in the desert west of Cairo and the Pyramids) found that the air there is purer, and the place quieter. They are no longer near factories and workshops, and there is room around the houses, and lots of green space. On the other hand, many services are not yet in place. The local councils are described as responsive to complaints and suggestions. The sewage system works. Those who had moved to Saqr Koreish (in the desert east of Cairo) found a similar situation, but without paved roads, even though traffic is heavy. One man moved to Dar al-Salam only because his house collapsed, but he would rather have stayed or moved to Sixth of October City. He found Dar al-Salam a step downward; "it is a slum," with lack of planning, congestion, workshops next to residences, lack of services and infrastructure, noise, and air pollution. The community is heterogeneous, not like the homogeneous one in Sixth of October City, and because of this, people have a hard time finding common grounds for cooperative action. He found the Dar al-Salam local government to be uninterested and ineffective.

We asked people in 1997 to compare their neighborhood to others.[31] Those who thought that there was a better neighborhood somewhere other than their own were 60.6%, ranging from 80% in Kafr al-Elow and 74.2%

in Dar al-Salam to 59.8% in Sayyida Zeinab and finally to only 22.3% in Abkhas. Men, at 61.4%, were slightly more likely to think this than women, at 59.8%. Asked to name such a neighborhood, the places people identified (in declining order) were: Ma'adi (mostly from Dar al-Salam), Helwan (almost all from Kafr al-Elow), new cities in general (mostly Kafr al-Elow), Zamalek (Sayyida Zeinab), Heliopolis (Sayyida Zeinab), Madinet Nasr (Sayyida Zeinab), Bagur town (Abkhas), Mohandiseen (Sayyida Zeinab). The overall pattern is that people chose a higher status neighborhood near themselves to which they would like to aspire. We then asked why it was cleaner: the answers included that it was a residential area, with no factories, garbage, etc., that the inhabitants are responsible (i.e., that these neighborhoods are inhabited by people with proper behavior who do not throw garbage or water in the streets, etc.), that the water is clean, that it has no noise, that it is a high socioeconomic area inhabited by better-educated people and has embassies, that it has a lot of greenery, that it gets attention from the government, or that it is well planned.

Conversely we asked if people knew of a worse neighborhood than their own. Fewer answered "yes" here: 30.3% on average, with more such answers from Dar al-Salam (41.1%) and Sayyida Zeinab (40.6%) than from Kafr al-Elow (25.1%) or Abkhas (11.7%). Men were more likely than women to be able to think of a worse area (34.5% to 26%), as they were of a better one. In other words, the people of Abkhas and Kafr al-Elow were more likely to see themselves at the bottom of the scale (they could not think of a worse neighborhood) than those of Dar al-Salam and Sayyida Zeinab. Those from Abkhas who thought of a worse place mentioned industrial areas, while from Kafr al-Elow the people mentioned the nearby area of 'Arab Kafr al-Elow. Both Kafr al-Elow and Dar al-Salam mentioned al-Ma'asara; both Dar al-Salam and Sayyida Zeinab mentioned al-Basatin, Bulaq, and al-Madbah. Sayyida Zeinab respondents mentioned Dar al-Salam more than Dar al-Salam respondents mentioned Sayyida Zeinab (33 to 5). Dar al-Salam respondents mentioned some of the poorer parts of Dar al-Salam, and Sayyida Zeinab respondents some of the poorer parts of Sayyida Zeinab. The reasons why these neighborhoods are worse include: they contain factories or informal squatter areas, are too crowded and noisy, offer a low standard of living, their markets, the presence of garbage and of sewage overflows, and negligence on the part of government and people.[32]

Street Food

Many in Egypt consider eating food prepared and sold in the streets a risk on the grounds of hygiene. Yet equally many do eat such food. In the overall

1997 sample, 57.6% said that they eat food prepared in the streets. To eat street food is presumably a sign on the one hand that people have a modest amount of discretionary income, and on the other that they are part of an urban lifestyle where families no longer automatically gather for meals.[33] Those who eat street food are a bare minority only in Abkhas; otherwise they are evenly spread. Men are more likely to admit this than women (64.2% of men to 50.7% of women). Among those who eat such food, 91.6% say they feel no risk. Those who do believe there is a risk mostly cite generalities: the food is not clean, it is not healthy, or they do not know how it is prepared.

Domestic Conditions

Critical in respondents' appreciation of their quality of life is the cleanliness of the streets and the general area where they live. For our analysis we focused on garbage, sewage, water, air, and noise, with some reference to other problems, such as insects and rodents that derive from the unclean environment.

Garbage (Solid Waste Disposal)

Garbage disposal is a major problem in Egypt. Two major issues in the private collection of garbage are the fees[34] for the service and its regularity. Some garbage is collected and removed from the Cairo neighborhoods, but much is not.[35] In the absence of regular garbage collection, the preferred manner of disposal is to put the garbage in a plastic bag and throw it on an informal dump (*kharaba*) at some distance from one's residence, trying to make sure that no one sees this act. There is nothing casual or thoughtless about this activity; it is planned and organized. If these dumps are not collected by the municipal workers, they are likely to be burned. Respondents recognize that burning garbage contributes to air pollution and is undesirable.

Overall in the 1997 sample (Table 7), 52.3% of the respondents say that a garbage collector comes to their house. This is almost universal in Sayyida Zeinab (99.2%) and in Dar al-Salam (98%),[36] and almost absent in Kafr al-Elow (3.7%) and in Abkhas (0.4%). The garbage collectors are predominantly private (70.4%, ranging from 79% in Sayyida Zeinab and 68.2% in Kafr al-Elow to 61.8% in Dar al-Salam). Only two cases in Abkhas reported garbage collection, both private. Most commonly the collectors pass every other day (73.6% of the cases) or daily (17.4%), and they generally are paid between LE1 and LE3 per month (91.9%).

Those in Kafr al-Elow and Abkhas who do not have a regular garbage collection service predominantly throw their garbage in the canal (59.1%), while another 16.7% burn the garbage, 10.5% throw it in the street (or in dumps in vacant lots), and 5% (in Abkhas and Kafr al-Elow) use it in the oven. Those who throw their garbage in the canal are 75.3% of the total in Kafr al-Elow and 42.3% in Abkhas, while those who burn it are 34.3% in Abkhas and 1.9% in Kafr al-Elow. In Abkhas 2% say they throw their garbage in the well. In Kafr al-Elow 16.1% say they throw their garbage in the street.

Among those who do not have regular garbage collection, women are quite a bit more likely than men to say they throw their garbage in the canal (67.2% of 548 women v. 51.1% of 552 men); while men are more likely to say that they burn the garbage (22.6% v. 10.8%).

Here it may help to describe briefly the system by which household waste was collected in Cairo at the time of the study. The garbage collectors (*zabbalin*) live in a series of settlements around the periphery of Cairo. They use donkey carts and small trucks to collect garbage from residential and other areas of Cairo, hauling the waste back to their settlements. There the waste is sorted, and the organic waste is fed primarily to pigs, goats, and other animals. The pigs are eventually sold to pork butchers in Egypt. The inorganic waste is then further sorted into metals, plastics, bones, etc., for sale to dealers in these products. Because their main income comes from the sale of pigs and recycled materials, the *zabbalin* are mainly interested in the "content" of what they collect, rather than in the act of collecting. Hence, this system often does not work very well in poor neighborhoods.[37]

Garbage collection is difficult in Dar al-Salam because private collectors rarely come in, and often find the narrow and uneven streets impassable. Some of the private garbage collectors are recruited by the local district government; others make deals with residents. Most are small operators so there need to be many to cover the district. They charge LE2–3 per apartment per month. Residents must also supply the plastic garbage bags. Some people prefer the government collectors since they are better equipped (pickup trucks instead of carts) and more reliable; others denied even the existence of public collectors. There are often quarrels with private collectors when they come late or refuse to haul away all the garbage. The collectors sometimes select what they can recycle, and leave the rest, or dump it around the corner. The garbage collectors themselves complain that there is little of value in Dar al-Salam garbage, and that the people are demanding of them: "They treat us as if they have bought us with their LE2." On the other hand, the collectors in Manshiyet al-Sadat and in Maʿadi are happy because the garbage from those areas is rich in recyclables.

In Sayyida Zeinab, too, it has proven difficult to get garbage collectors to service the area regularly, and street sweepers are rare and inefficient. Some of the worst problems are in the alleys and back streets referred to as *zarayeb mansiya* (forgotten stables), where there are problems of garbage dumping, unpaved roads, noise, and air pollution. This contrasts with the main streets where private shopkeepers help maintain cleanliness.

The many garbage dumps in Dar al-Salam and elsewhere are found in empty lots and similar sites. The dumps encourage mosquitoes, flies, and rodents. One pile of garbage our team examined included food, broken pens, plastic boxes, cigarette butts, paper, bones, and other items. In a poorer area like Old Dar al-Salam, the garbage does not contain any food or bones. Usually when the garbage is just dumped, it is dumped away from the house. A resident of Dar al-Salam noted, "People use unused land to dump their garbage. They usually prefer to go to a site away from their own house. It is a very early morning mission, carrying the garbage of the day before in a plastic bag, avoiding being seen by people who live next to the dump site in order not to have a fight that might end in carrying the bag away and looking for another dump site." In Sayyida Zeinab we heard that people sometimes throw garbage from their windows, though they are careful not to do so when someone may be watching this "bad behavior."

A fruitseller near an empty lot on the edge of Dar al-Salam noted that people come in the early morning to a spot near his stand to throw away their garbage. He has had fights with them, and he has tried to stop people from doing this, but there are still many who do it when he can't see them. This young man is originally from Asyut, and said he never thought that people in Cairo were so dirty. He added, "Sometimes I feel so embarrassed when a very respectable looking man is throwing a plastic bag of garbage on that dump. I find it difficult to tell him that he shouldn't be doing that." The fruitseller himself burned his leftovers, knowing it was an imperfect solution because of the smoke, but the best available to him.

There are some public garbage containers on the streets, but they are not effective. For one thing, they are poorly maintained and seldom emptied, and so they end up with the same problems as any dump. Not all the garbage gets into the dumpster, some littering the ground around it.[38] Women try to keep their own doorways clean, as well as their houses, but do not tackle the rest.

Burning garbage is common, and this is considered a separate problem. Some women, in Kafr al-Elow and Dar al-Salam, argued that the fumes and the smell produced by burning garbage are worse than just throwing the garbage away, while others preferred burning. Some women felt that burning garbage produces all sorts of allergies and chest diseases, in addition to the bad smell. Burning plastic is regarded as particularly problematic.

Markets in Kafr al-Elow, Dar al-Salam, and Sayyida Zeinab are well recognized as sources of garbage. Some of the previous day's garbage produced by the vegetable market under the Dar al-Salam bridge is hauled away by the municipality in the morning. The garbage produced by the market in al-Fath Street, Dar al-Salam, is disposed of as follows: spoiled fruit and vegetables are sold to women to feed their poultry; hay, baskets, and papers are burned at the end of the day; the remainder is collected by the municipality at the end of the day for LE3 monthly per trader.

The annual *moulid* (festival) of the Sayyida Zeinab mosque adds to the problem of garbage in particular, as well as noise. Many of those who attend the *moulid* come from rural areas (Othman 1997; Abu-Zahra 1997). The *moulid* is also a time when there are crowds on the streets, and some youth hassle other folk: moral pollution.

Waste from clinics and dispensaries is also collected by the garbage collectors, after it is put in a box, can, or plastic bag. Some clinics have no regular garbage pick-up ("because they have so little") and instead throw their garbage in a public waste bin, or in any other garbage dump.

In Abkhas, people bemoan the absence of an organization to remove solid waste from the village. Domestic waste is sometimes dumped in the Bahr Shebin canal—garbage, dirty water, excrement of animals and birds, etc. There is a garbage disposal area on the banks of the canal and although the area is plowed and cleaned by tractors, people still dump their garbage on the banks of the canal. Some therefore feel a guard is needed to prevent this, and that people who continue to throw their garbage there should be fined. Generally people feel that it is hard to stop people dumping in the river because it is public. There are also other garbage disposal areas. Alternative disposal systems for garbage are absent. There is no easy solution.

In the old days in rural and semirural settings like Kafr al-Elow and Abkhas, people used to combine garbage with hay or straw to create litter for cattle, or they would burn waste in the ovens to bake bread. But modernity has reduced these activities. Also, the amount of garbage has increased with the population. The garbage problem is now more severe because people use more paper, and bake in a modern oven. With no system of garbage collection in Kafr al-Elow or Abkhas, people dump their household garbage in the canal. The poorly discarded garbage and sullage provides a suitable environment for flies and mosquitoes to thrive, which people find to be a major irritant. People block the ventilation in their houses to escape the dust, and then when they are bothered by insects, they use insecticides in this closed place, leading to further problems.

Sewage

In Sayyida Zeinab, an old urbanized area, there was a sewage system under renewal at the time of research. The sewage network was being extended in Dar al-Salam, so some areas had connections, others not. There was some sewage in Kafr al-Elow, but none in Abkhas. Houses that were not connected to sewage systems generally had a *transh*, a cement cesspit or sewage tank (A. Nadim et al. 1980:96). These cesspits had to be emptied out periodically. In Abkhas there were usually unlined latrines, where much of the fluid passed into the soil.

If getting clean water in is a problem, then so is getting dirty water out. In 1995, 49.3% of the buildings in our sample were connected to the sewage system, but there was a considerable range from Sayyida Zeinab, where everyone was connected, to Abkhas, where no one was. Two thirds of the buildings in Dar al-Salam (67%) were connected, but only one fifth of those in Kafr al-Elow were (19.5%). Toilets in the household were also nearly universal. Overall, 97.4% of our households had them, with a range from 93.3% in Kafr al-Elow to 99.4% in Abkhas. Two thirds of the respondents (66.7%) said that the family laundry was done inside the house, either in the kitchen or the bathroom (so the excess water had to be discarded through the sewage system). The range here was from 94.8% in Sayyida Zeinab to 24% in Abkhas, with Dar al-Salam at 83.8% and Kafr al-Elow at 54.9%.

In the late 1970s, when roughly a quarter of Cairo's population were without a potable water supply, Asaad Nadim and colleagues (1980) surveyed several Cairo neighborhoods that lacked potable water and sewers. Some of these deprived areas were poor enclaves in otherwise equipped areas, while others were on the "rurban fringe" (rural/urban). About half the households supplied themselves with water from public taps, where waiting and quarreling were common. Another quarter took water from neighboring buildings (p. 77). In these areas, 94% of the households had "private toilets of some kind," and 72% used tanks to collect toilet and other waste, while about one fifth relied on the public sewer system. Other waste water was dumped in the street or a nearby canal, with some care taken not to disturb neighbors (pp. 95–105). In Manshiyet Nasser in the 1980s there were "rules" to prevent pools of standing water in the streets: "Neighbors insist that women do not dump their water at a single point, but sprinkle it around in the vicinity of the front door, to minimize the formation of ruts on street surfaces" (Tekçe et al. 1994:38). Women in rural areas were the managers of waste water, and were careful not to overfill the tanks and not to dump too much water in the street (El Katsha et al. 1989:37).

In 1997, 99% of houses in Dar al-Salam and Sayyida Zeinab were connected to sewage (all but four cases out of 1,200), and 58.2% in Kafr al-Elow were. The sewage system had been extended in both Dar al-Salam and Kafr al-Elow between 1995 and 1997, so that sewage connections became practically universal in Dar al-Salam and rose from one fifth to three fifths in Kafr al-Elow. In Abkhas, 99.6% of houses were not connected to sewage.

About one third of the respondents (37.1%) said they had paid for their connection. This figure was highest for Kafr al-Elow (82.9% of those with connections), then for Dar al-Salam (44.2%) and lowest for Sayyida Zeinab (3.2%). Probably the figure for Sayyida Zeinab reflects the age of the system, which may be older than our respondents. Thus, the payers are mostly in Kafr al-Elow and Dar al-Salam, which have had the most recent expansion. Most people who had paid said they did not know how much they had paid for the connection; of the others, 60.9% said they had paid over LE100. Of the 575 who had paid, 39.5%, mostly in Dar al-Salam, said they had paid more than two years before, while 200 (34.8%), mostly in Kafr al-Elow, said they had paid less than a year before.

Of those with sewage connections (N=1,548), 12.1% said they paid for sewage maintenance. These people were found in the three urban neighborhoods, but mostly in Kafr al-Elow, and least in Sayyida Zeinab. Those who paid mostly claimed to pay less than LE5 a month, while a few said they paid more than LE15 a month; the payment is typically either once a month or whenever the system is clogged.

Almost all of those without sewage connections said they had a cesspit or tank. Only 22 households claimed neither a sewage connection nor a cesspit; 17 of these were in Abkhas, three in Sayyida Zeinab, and two in Kafr al-Elow; 15 were female respondents. The cesspits are infrequently emptied: most respondents said that they were evacuated less often than once a year (37.2% of those with cesspits), or that they were never evacuated (28.9%). Since there are few cesspits in Sayyida Zeinab and Dar al-Salam, the contrast is between the other two sites. In Kafr al-Elow the pits are emptied more frequently than in Abkhas: 64.3% of pits in Kafr al-Elow are emptied more often than once a year, compared to 1.4% in Abkhas; this difference reflects the prevalence of lined and unlined pits. The cost ranges from less than LE10 to more than LE70, with 51% paying LE30 or less. (In Abkhas some of those who said they never emptied out their pit nevertheless gave a cost.) The task of emptying out the pits falls either on the local council, especially in Abkhas, or on private individuals, especially in Kafr al-Elow. Of the 530 cases on which we have information, 40.8% were emptied by the local council, 41.9% by a private contractor, and 15.8%, mostly in Abkhas, by the householder.[39]

Old Dar al-Salam was connected to the sewage system in the 1960s, al-

Mal'a in the 1970s, Manshiyet al-Sadat in 1987 (with the addition of booster pumps in 1994), al-Gezira at the time of research in 1995–97. At the start of the research, only al-Gezira and al-'Ubur were not connected, and depended instead on lined underground tanks. With this system women would discard used water in the streets in front of their homes rather than add to the tanks. When we were first reconnoitering the area prior to the research, women would call to us from the balconies to encourage us to bring a sewage system even though they did not know our reason for being there.

The tanks were typically emptied every week. Most people who emptied the tanks were private operators. As soon as they appeared, the municipality stopped providing this service. It was considered a good investment for these private operators. In this area about 25 wagons did the work, and about 70 people were employed, though with a fairly high employee turnover. Each wagon required a driver, and an assistant to put the hose into the tank and to start the suction pump.[40] Usually the operators were not the owners. The operators' costs included payment of bribes to traffic police to permit them to empty out their vehicles. A favorite dumping place from Dar al-Salam was a sewer in al-Basatin where a local tough collected LE5 per day for each wagon. Those who did not want to pay LE5 emptied their wagons in the empty land between Al-Matba'a and the Ma'adi military hospital. In Kafr al-Elow, the wagons were emptied into the canal. The Nile police and the municipality would arrest those who attempted to drain their wagons into the Nile itself. There were also brokers or agents who informed the wagon operators when someone needed to have the tank emptied. In early 1996, just before the completion of the new sewage system, the price reached LE14 for one evacuation, of which LE2 went to the broker as a commission. The cost of emptying the tank was then shared among the residents in the building. The operators were conscious of providing a necessary service. One said, "Thanks to us, al-Gezira is not drowning in sewage water," as it nearly did when the operators went on strike in 1991.

With the installation of the new sewer system, the operators have transferred their wagons to other places, such as Ma'asara or other communities on the edges of Cairo, that still lack a sewage system. However, there is still a niche for some wagons, since the sewers are occasionally blocked and overflow.

When the new sewer system was installed by the government in al-Gezira/Dar al-Salam, people were satisfied (see Tadros 1996). Then they discovered that they had to pay for their personal connections, though the main network had been paid for by the government. In some places in Manshiyet al-Sadat and al-Mal'a, private contractors tried to connect buildings to the sewage system, but did a poor job so that blocked pipes often flood the streets. Each building was assessed and had to pay a certain sum, depending on the number of apartments. The amounts came out higher than people had anticipated: Typically a two-room apartment paid LE50, and a three-room apartment LE80, and double for shops. However, in the end people came around when they remembered what the alternative was, without a sewage disposal system. "The government has been boasting for quite a while about the sewage project, and now we are paying for it," was one comment. People also observed that since the new sewage system has improved conditions in the streets, there are fewer mosquitoes, and property values have risen. They have also noted that a good sewage system increases water consumption. Constant construction, on the other hand, keeps the streets torn up. Now that the sewage system has been built, people are reflecting on their next collective goal (perhaps paving the streets).

There is no sewage disposal system in Abkhas village. Each house has its own unlined latrine that the owners clean themselves or hire someone to clean for them (sometimes using buckets). Generally the sewage is transported to the fields and used there as fertilizer. It is covered with ashes to make manure. The tanks fill up very slowly, in one reported case taking over 25 years. The cost of cleaning them is estimated at LE50.

Water

Virtually all households in the three urban sites had access to piped water through a tap either in the household itself or in the building. In our 1995 sample, 85.1% reported living in a building connected to running water. The three Cairo sites ranged from 95.2% connected in Kafr al-Elow to 100% connected in Sayyida Zeinab, while Abkhas was much lower at 35.3%. Two years later, in the 1997 sample (see Table 7), 86.7% of respondents said their household was connected to the water system.[41] The unconnected were still mainly in Abkhas, where they were 56.1% of the total (i.e., 43.9% were connected, so there had been expansion since 1995). In addition to the 284 households in Abkhas, there were 20 in Kafr al-Elow and three in Dar al-Salam who were not connected. The connections were overwhelmingly through the government system (97%), with most of the others citing pumps. Some of these were in Dar al-Salam (18 cases) and Sayyida Zeinab (20 cases) in addition to Abkhas (10 cases).

The urban sites overwhelmingly use the government system for their drinking water (95.3% to 96.5%), while Abkhas overwhelmingly uses pumps (98.8%). It seems that the people in Dar al-Salam and Sayyida Zeinab who have pumps use them for drinking water. Some 79.6% of our respondents say that the water they drink is good for drinking, with the greatest level of satisfaction in Abkhas (90.7%). In the three urban sites about a quarter of the respondents were unhappy with their supply from the government system. The objections to the water are largely that it is not (visibly) clean, being turbid or dirty, or that it is not clean because it contains chlorine or other chemicals. Some in Kafr al-Elow (n=31) thought that there was a color change in the water because of aluminum. Some in the three urban sites were also not satisfied with the taste, mostly because it is "different," bad, or bitter. Often the key taste test involves what the water does to the taste of tea. Some also objected to the smell of the water, and in Dar al-Salam, some 22 people pointed out that the water tastes of sewage. Perhaps some of these were among the people (n=25) in Dar al-Salam who found solid waste in the water. Some 169 respondents, with the largest group in Sayyida Zeinab, noted that the water contains cement, dust, salt, suspended matter, etc.

Overall in the 1997 sample, 170 people (7.4% of the total sample) said they treated the water before drinking it. The largest group was in Dar al-Salam (n=64 or 10.6%), then Sayyida Zeinab (n=54 or 9%) and Kafr al-Elow (n=47 or 7.8%), finally Abkhas (n=5 or 1%). Of those who treated water, 77 boiled the water, 41 filtered it, 31 let it sit until it clears, 13 put it in porous ceramic water jugs or filtered it through cotton, six put it in the refrigerator, and two gave no answer. Filtering is more common in Sayyida Zeinab, and boiling in Kafr al-Elow and Dar al-Salam. Of the five in Abkhas who do something, three simply put the water in the refrigerator. Essentially the same picture is true for water for cooking.[42]

When it comes to water for washing, some Abkhas residents switch to the government system or to the canal (pump, 43.1%; canal, 35%; government system, 21.1%). The people of Abkhas also use pump water (50.6%) or government water (39.7%) for bathing, but largely avoid canal water (8.7%). The urban sites also predominantly use government water for washing and for bathing.

Participants in the focus groups in the three urban locales also complained that the taste or smell of the drinking water meant that it is polluted. In Kafr al-Elow, they linked this to Nile pollution, in Dar al-Salam to defective pipes infiltrated by sewage, and in Sayyida Zeinab to the lack of pressure in the distribution system. One enterprising young man in Kafr al-Elow reported an experiment: he placed a sponge into a glass of drinking water for a week; it turned black, indicating to him the high level of contamination.

Attitudes towards water sources: Abkhas

Water issues dominate the environmental concerns of the people of Abkhas. People cite the pollution of the waters of the main canal, the Bahr Shebin, that passes by the village, and they also worry about the quality of their other sources of water—shallow wells with pumps, and the government system of a deep well distributed by pipes to household taps. Water is evaluated according to color, smell, taste, and the presence of visible particles (turbidity). Some people thought that the canals could not be polluted because the water in them is flowing, or at any rate that running canal water is cleaner: "it doesn't retain the garbage." But other people rejected this idea. Another water problem is the rising level of ground water.

The main source of pollution of the canals is people dumping solid waste (the LE1,000 fine for dumping is not enforced) and emptying out sullage tanks (the tanks are drained by carts that empty into canals or waste land areas). Disposal of plastic is cited as a particular problem. At the same time, people feel that there are no viable alternative ways of disposing of waste. Water is contaminated by dead animals, household waste, and sewage water. One woman said her feet were cut by broken glass. Another noted that the fish are disappearing. Women worry about the pollution of the canal water because of all the uses people make of it, from washing and fishing to bathing or swimming. Some argue that washing aluminum utensils in canal water can lead to kidney failure. Young men noted that some of the canal pollution came from industries upstream.

Some farmers argued that the slowing down of the flow of canal/river water by dams leads to a reduction in water quality. Some mentioned that before the construction of the High Dam, the river used to be red (i.e., with silt) and it would flood. The silt also helped increase yields. Another problem with canal water is the prevalence of "worms and bacteria" leading to diseases like schistosomiasis. Some others pointed out that polluted water was harmful to irrigated crops, or to animals drinking from the canals. There are still some who prefer to drink canal water despite the problems, though the number is diminishing.

Village leaders, all men, thought that the problem of dumping was caused by lack of awareness on the part of village women, who dispose of their household waste by burning it or by throwing it into the canal or dumping it on the banks of the canal. Village youth (male) are also partly to blame.

Piped water is often not available, especially for people on the third floor or above, and it is considered to be dirty and suitable only for washing clothes, not for drinking. One reason for this is rusty pipes; another is that chlorine is not added to the water, although adding chlorine degrades the taste. Water storage tanks (*sahareeg*) are not clean. Piped water smells and tastes bad, and some think it contains sediment, mineral salts, and worms.

Pump water is preferred for drinking, and for watering livestock,[43] though some feel that there is a real danger that this water might be mixed with sewage and cesspit overflows, especially if there is a cesspit near the well. (Whether this was really a danger was a controversial point among participants.) It is therefore thought important to dig the well deep enough to tap sources below sewage and cesspit overflows, and to locate it carefully. In an interview, one person said that water from the pumps is best because it comes "from the seventh layer of earth."

In an interview, a woman compared the water from different sources, "In our houses we receive water from the pipes, yet I don't use it for food. Even when I make tea from it, it looks muddy. I use it only for prayer ablutions and sometimes for washing up. Other times I do the washing up from the water pumps or the river [canal] water. I also wash the kitchen utensils in the river since the water is clean and makes the pots and pans shiny clean. As for tap water it makes the kitchen utensils discolored and dull."

Another argued, "Generally speaking, the river is clean yet the people are not. For instance, while a woman is washing, others would come and wash off their dirt onto her laundry. Isn't it disgusting that while I'm washing kitchen utensils that we eat in, another woman would show up and start washing trousers or children's diapers? Nevertheless, the river water doesn't retain the dirt."

And a third noted, "River water is flowing and not stagnant, therefore it is not contaminated by any diseases. The diseases are found in the small canals, that is where you find bilharzia."

Water disposal can be a problem, too, and one of the consequences is that women often prefer to wash clothes and dishes at the canal in order not to have to dispose of dirty water. This also has the advantages that the task is more sociable and that the softer canal water requires less soap.

Air

Participants were also sensitive to the amount of dust (*turab*), in the environment (i.e., perceptible matter in the air). Over two thirds (69.8%) of our 1997 respondents found that their neighborhoods were more dusty than they used to be (Table 7). This feeling was strongest, indeed almost universal, in Kafr al-Elow (98.3%), followed by Abkhas (70.6%; note that the village is second highest), while the figure was just over half in Sayyida Zeinab (55.3%) and in Dar al-Salam (55.1%). The most common solution to the dust problem was to put screens on windows (25.6%) or even plastic sheets (23.4%). Others mentioned shutting the windows (10.6%) or adding curtains (9.4%). Only a few (1.8%, mostly in Kafr al-Elow) mentioned correcting the problem at the source. These people argued that factories should install filters, that the government should do something about the factory (such as moving it), or that people should complain to or about the factories. At the other extreme, 20.9% of the total sample (n=2,307) said that they could do nothing. These cases were most numerous in Sayyida Zeinab and Dar al-Salam, where also the perception of dust was lowest. In Sayyida Zeinab this option of "nothing" was the most common choice (37.9% of 599 cases), while in the other localities the most common choices had to do with sealing off the dwelling by blocking the windows.

Air pollution is the highest level of concern in Kafr al-Elow. Many focus group participants in Kafr al-Elow say that they are surrounded by three cement factories, and they are concerned about cement dust from them. Cement dust is mentioned first and in most detail by all age and gender groups. The people who live in Kafr al-Elow experience the dust, but also a certain number of them work in the cement factory and that gives them an additional source of information. The problem of cement dust is linked to the ongoing debate about the presence and use of filters in the factory to capture the dust before it leaves the exhaust chimneys. The people feel the filters are effective when used (they estimate the dust is reduced by 60%[44]); the problem is that they are not used consistently.

The focus group of workers developed in greater detail themes that were present in all focus groups. A group of factory workers explained that filters are present, but that there are problems in operating them.[45] One problem is the lack of trained and skilled workers, so the filters are not used properly. Sometimes the workers forget to operate the filters because they fall asleep on the job. At other times, the filters break down under conditions of high pressure and high voltage, or because of careless acts by workers, or because they require cleaning. Since the maintenance staff does not work at night, breakdowns are dealt with by unspecialized workers with uncertain results, or have to wait until the following day. At times repairs take up to a month, during which time the factory continues to operate.[46] Another theory is that the engineers close down the filters overnight to increase production. Residents in Kafr al-Elow feel that the worst deposits of cement dust are at night or early in the morning. Some of these comments by workers blame the workers rather than the technology or the management. One comment sums up some of these attitudes with regard to the engineers: "The one who stops the filters in order to increase production cannot be following religion."

In addition to the ineffectiveness of the filters, subject to breakdowns, there is the problem of the disposal of the dust collected. This is supposed to be trucked back to the desert areas from which the raw material comes, about 15 km away, but there are not enough vehicles, and it is costly. Also, this area is gradually reaching its maximum capacity to absorb waste. So far, the workers maintain, efforts to find a constructive use for this waste have not succeeded.[47] In one focus group of men this waste was estimated at 16% of total production.[48]

One of the group of workers maintained that "If the company's management could politically afford it, they would have stopped the filters altogether, saved on the cost of the trucks for cement waste disposal, and also obtained higher production. But that option would be murderous to the area, given the amount of pollution that would be generated." One somewhat paradoxical suggestion is that although the managers oppose the use of the filters they still benefit from the commissions they receive to allow the installation of the filters. The workers also quote a story that a new method for blocking cement dust has been invented, but that the managers are refusing to consider it because they are corrupt.

Another issue is the change in the manufacturing process that the Turah Cement Company, and presumably the other cement companies, carried out around 1980. At that time, the company switched from a "wet" method to a "dry" method, involving new kinds of ovens and a different technique.[49] This technique aggravates the pollution problem. The "wet" method involved adding water to create a paste, and this did not generate the kind of pollution of gases or dust that the dry method does. However, the dry method is said to be less expensive in terms of production costs. Over the years, the factory has been expanded, with increased production and hence increased cement dust. Thus "the residents of Kafr al-Elow feel the gravity of the problem much more than in the past."

The workers are also conscious of the unhealthy working conditions in the factory, though they note that the factory maintains a health plan for its workers who undergo regular checkups. The managers of the factory are less affected than the workers because they seal themselves off within closed, air-conditioned offices. Workers sometimes knowingly undertake hazardous jobs because of the higher salary and bonuses, but they would try to prevent their children from doing the same. Some workers feel that the use of filters would reduce production and hence their own salary and bonuses. The group of young men noted that the health and welfare of the people were secondary to the immediate financial gains of the workers (through bonuses), increased levels of production for the management, or more production and more revenue from export and sale for the government.

The workers were conscious of the impending privatization of the factory (although there was no plan and no date at that time). They felt that the public sector is being allowed to run down, and that neither it nor the people who are part of it count for much. The factory has been cutting down its labor force, so there was a labor shortage in the factory. Even people willing to overlook the health hazards could not find jobs. "Now the only thing that counts is the private sector, money and people with connections. Since Kafr al-Elow and the workers in the cement factory lack those things, their complaints and lives are inconsequential." Other participants in Kafr al-Elow hoped that private management would be more sensitive to community pressure. Nevertheless, some people filed a suit against the factory and sought to attract the attention of political authorities up to the prime minister, but so far without result. A common feeling was that the people of Kafr al-Elow were insignificant compared to the elite living in Ma'adi and similar areas, who have more clout.[50] "The problem with the government is that they are not interested in the welfare of the people and the area, and the residents have nothing to remember but unfulfilled promises that date back to the times when they were born."

Apart from using the filters or other methods to clean up the waste from the cement factory, the only other solution is to move the factory to the desert. This is a long-term solution. On the other hand, individuals can move from Kafr al-Elow. Changing residence is locally recognized as a solution, but one not available to all because of the general poverty of the area. Also the most dynamic individuals move away, leaving those unwilling or unable to do something about the problem. Some of the young men suggested appealing to WHO to monitor the water and air pollution problems.[51]

Focus group participants in other localities also cited air pollution. Men in Dar al-Salam saw the source of air pollution in street dust, workshops (automobile paint, sawdust), and the burning of garbage. Women added car exhaust to this list. Most participants in Sayyida Zeinab attributed air pollution to vehicle exhaust. Vehicle exhaust has negative health consequences for people, especially children, and adds to the already high levels of lead in the air they breathe (and they argue that the government should introduce unleaded gasoline). The government's own vehicles, such as city and factory buses, are a major contributor. Air pollution is aggravated by construction or by unpaved roads, and pollution from construction is aggravated by the lack of coordination between the different government departments involved. Some note that the air pollution is not as bad in Sayyida Zeinab as in industrial areas. Both in Dar al-Salam and in Sayyida Zeinab, participants noted that air pollution contaminates the food that is publicly displayed for sale.

We interviewed a number of people in Dar al-Salam on what they knew of lead. The answers show quite a wide spread, but generally reveal that our respondents were not familiar with the idea of lead as a risky pollutant in the air and through other media. One of the better informed answers came from a worker who said, "Lead is a material that exists in lots of stuff, the most of important of which is car exhaust. This pollutes the air and is also a danger for public health, especially for children and pregnant women." Several others noted that lead passes from car exhaust to the air, but were not clear on the health consequences, thinking, for instance, that it affected the lungs. Two identified newspapers and bread sold on the street as sources of lead. For instance, a house painter noted: "The thing that is really full of lead is newspapers and that is why they no longer use them to wrap food. Lead is in the ink, and can gradually increase in the human body and can eventually cause poisoning." Several others denied the risk from atmospheric lead. Tangentially a housewife commented, "We were once incinerating the garbage in the street and one of our neighbors told us that the smoke is full of lead and extremely dangerous. We stopped for a while, then we resumed. What can we do?"

Noise

> Noise is my worst problem. Every day and all through the morning till the afternoon we have to live with the noise that comes from the machines used in the sewage project. This is unbearable. If they just finish this project it would be a blessing.
>
> Teenage boy, Imbaba, 1993

Excessive noise (to paraphrase Mary Douglas, "sound out of place") is another problem that affects the quality of life in Egypt. Noise is a health hazard and interferes in communication. Exposure to continuous or intermittent high noise levels may result in annoyance, stress, fatigue, and loss of hearing.

Overall a large minority (40%) found that their neighborhood was noisy (Table 7), of whom 92% said that they were disturbed by the noise. Women were more likely than men to see their neighborhood as noisy (42.3% v. 37.8%). The perception of noise was highest in Sayyida Zeinab (59.1%), followed by Dar al-Salam (54.4%), then Kafr al-Elow (36.3%), and finally Abkhas, where only 4.7% found noise a problem.[52] Of the people who noted a cause of noise (n=1,348), the largest number indicated children playing and neighbors (30.1%), to which might be added the 9.4% who cited "too many people." Then, in descending order, were: metal workshops (18.8%), traffic (17.8%), and carpentry workshops (4.5%). Coffee shops and vendors were mentioned by 5.8%, the sound of televisions, cassettes, and microphones by 5.1%, and the marketplace, shops, and pigeon houses by 4.6%.[53]

In the Dar al-Salam and Sayyida Zeinab focus groups, participants complained about noise caused by playing cassettes loudly, by workshops (e.g., car repair), by coffee shops, and from other sources. Traffic congestion adds to the noise. The workshops in Sayyida Zeinab operate long hours, including late at night. This is the area in which some (senior women) complained about workshops not only because of the noise engendered by the work but because of the filthy language ("words out of place") used by the workers. In Dar al-Salam, participants complained of noise from children and youth playing in the streets, coffee shops, and tape recorders. Main streets and markets were considered a major source of noise. Those living near a bakery find that is a source of noise because of the quarrels between the bakers and their clients.

Other Problems

Pests

Pests are particularly a problem in Kafr al-Elow and in Dar al-Salam. The flies, mosquitoes, cockroaches, and rodents (rats and mice) flourish along canals used as dumps and in solid waste dumps. In Dar al-Salam, the new sewage system is helping to reduce their number, although they still exist because of the piles of garbage and the unpaved streets. The insect problem is worst in the summer. Many residents use screens on their windows to keep insects out, or use chemical insecticides inside.

Nearly two thirds of our 1997 sample (65.1%) said they used pesticides at home to spray for flies and mosquitoes. The figure was fairly even across the four localities, ranging from 61.9% in Sayyida Zeinab to 68.9% in Kafr al-Elow, and male and female respondents gave essentially the same answer. However, those who agreed that pesticides affect health were much higher. Overall the figure was 89%, with very little variation either by locality or by gender. Thus, many of those who use pesticides are aware of the potential threat to health that they represent, but presumably continue to use them because of the insect pests they confront. Also, 95.6% of our respondents felt that their dwellings had enough windows to provide air (from a low of 91.3% in Kafr al-Elow to a high of 99.4% in Abkhas), and about the same percentage felt that a lack of air could affect their health (94.7% overall).

Chemicals

Chemicals are used in Egyptian agriculture, primarily as fertilizers or pesticides. In general people in Abkhas are aware of the links between chemicals and health, between behavior regarding chemical use and well-being. They are alert to sources of pollution. They know at some level that the chemicals passing through their environment are potentially hazardous, and they keep some kind of eye on these chemicals. They recognize that some are more dangerous than others. For instance, they know that urea is a dangerous fertilizer, especially to livestock, and use it cautiously. After spraying, people feel they should wait 15 to 20 days before using anything from an adjoining field. Urea is not used on *bersim* (alfalfa) because people think it causes the cattle to lose blood in the form of red urine, and they eventually die from loss of blood. The "super" (superphosphate), on the other hand, is believed to be beneficial for cattle as it provides them with the calcium and phosphorus they need.

According to the head of the agricultural cooperative in Abkhas, those who spray cotton with pesticide wear protective clothing, and a representative committee from the cooperative with an ambulance is present in the fields with first-aid equipment. If one of the laborers forgets to wear a mask or if his hand is dirtied by the spray, they perform all the necessary first-aid procedures. They also use soap to wash their hands in order to eat their lunch. Smoking and eating are prohibited during spraying. Even if this account is idealized, it reflects an awareness of the problem.

Most vegetables are used unwashed in the households of Abkhas, as most of the village women say, "The vegetables coming from the fields don't need to be washed as they are already clean." The villagers believe that "eating from the fields is natural and clean since it hasn't been touched by any hands," thus implying that human contact is more hazardous than chemical contamination. However, vegetables such as okra, cabbage, and *mulukhia* are covered with dust and the exhaust of cars and motorcycles that travel along the main road. In addition, corn and cotton near the vegetables are sprayed with pesticides, some of which reaches the vegetables. There are reports from elsewhere in Egypt that pesticide is mixed with stored crops to protect them; this is then rinsed with water before the grain is ground (Mehanna, Hopkins, and Abdelmaksoud 1994:73).

Focus group participants in Abkhas noted that pesticides are a threat, but perhaps less so than in the recent past since their use has diminished as people have become aware of the dangers. Abkhas women report a lot of spraying for flies and mosquitoes, which they feel causes chest allergies. The young men note intensive use of pesticides and chemical fertilizers in the fields to increase agricultural production. Soil quality is deteriorating, and there has been a decrease in the use of organic fertilizers, so more chemical fertilizers are needed. The older men note that the "extensive use of chemicals in fertilizers as a means of increasing production is a prime source of disease," while at the same time noting that this did not deter people interested in profits. Community leaders noted that "The intense use of the internationally forbidden urea in fertilizers as well as other dangerous chemicals has a detrimental effect on soil fertility." Some stated that the more pesticides accumulate in the soil, the more fertilizers you have to use. The use of pesticides is risky for animals as well as for humans. Pesticide use in cotton cultivation can be reduced by using old-fashioned hand-control methods. Or the government can research and develop less dangerous fertilizers (rather than urea). The currently available alternative found in the cooperatives is not effective.

Moral Pollution

A lot of people in all localities stress moral pollution—for instance, the behavior of boys towards girls in the streets, interpersonal relations in general, or the use of drugs and *bango* (marijuana).[54] One definition of environment, in fact, is a person's origin and morals, or their "morals and behavior." One young man noted that moral degradation is an environmental problem: "All the kids are brought up in a dirty environment and that affects all their behavior toward the environment. And that in turn comes from their parents' education."

An example of the kind of situation some people find disagreeable is that in Sharia al-Nassiriyya in our Sayyida Zeinab area. This street contains numerous cafés, many of which include a video cassette player and TV sets to show films. Some have recently introduced video games. Most of the films shown are thrillers, horror films, or erotic films; the majority of the audience are craftsmen and students skipping school. There are frequent quarrels, both verbal and physical. Customers smoke, gamble at cards, and indulge in drugs. The street is often noisy. The street also serves as a headquarters for many groups of musicians who specialize in wedding processions (*zeffas*). When these groups gather, noise is likely, not least from arguments over wage distribution. Some people in Dar al-Salam also worry about drug dealers and addicts at night; they believe there are places (notably certain cafés) where men gather to use drugs and watch pornographic films. Young thugs try to prove their manhood by fighting. The relative absence of police means that people must settle affairs themselves, give in to thugs, or rely on the courts.

A woman from Imbaba, a mostly informal area in north-western Cairo, interviewed in the 1993 pilot study, complained, "How can I, a village girl, allow my children to grow up in such a community? It is not only the problem of pollution that bothers me, but the fact that nobody will care to help me when I am in trouble." Such concerns bring us back to Mary Douglas's observation that pollution is equivalent to disorder, including moral disorder.

Quality of Life in the Research Sites

Comparing our data to those presented by Nadim et al. (1980) and by Tekçe et al. (1994) suggests a gradual improvement in the quality of life has occurred, though it should be remembered that the questions and the neighborhoods were not the same. People in our sample report conditions that are generally crowded, noisy, and dirty, though sewer systems and other amenities are being extended. The rural locality differs in some key respects from the three urban ones (dwellings are newer and larger; there is less concern about health and the environment). There is concern about clean streets everywhere. There are also continuing problems, especially with water and waste removal. Despite the problems, people are relatively satisfied with their living conditions, including their neighborhood to which they are generally attached. In this chapter we have not only seen the conditions, but the strategies that people adopt toward their problems and some of their attitudes toward their living circumstances.

5

Cultural Construction of the Environment and Pollution

> Environment is the origin that you come from. My environment is my origin. For example, people do not agree to let their daughter marry a bad [man] from a bad origin, a bad environment. Therefore when you talk about environment you are really talking about the origins of men.
>
> Cairo taxi driver, 1993

In this chapter, we present material on the understanding and construction of environmental issues. Unless otherwise specified the data come from our first survey (1995). However, in our second survey (1997) we asked some additional questions to complete or sharpen the picture. Given the similarity of the samples we treat all the answers as having essentially the same epistemological status. We used some questions that asked for general definitions, and some that probed by asking people to compare or to rank different circumstances, or that provided us with the material to do so. The overall picture shows how respondents think about the environment, the choices they prefer, and the understandings they have. It reveals who is concerned, what they are concerned about, and what fears they may have. There is also additional material on how people interpret the causes of pollution and environmental change, and who or what they hold responsible for these changes.

Since we are dealing with a dynamic situation, people's sense of the rhythms of change is also germane. Some questions were asked on both survey rounds, and we can detect some short term trends. The final section of the chapter deals with the influence and role of the media, as our respondents saw it.

This chapter should be seen in light of the preceding chapter on living conditions and as leading up to the material on political action in chapter 7.

Definitions and General Attitudes

Similar answers to questions on environment and questions on pollution are given in the surveys and in the focus groups. In the common use of the word *bi'a* (normally translated as "environment") in Egypt, the physical and the social environment are confounded. In fact, pretty clearly some people understand primarily the social environment by the word. For instance, some cited the moral pollution of youth among environmental issues, and a group of men from Dar al-Salam cited customs and tradition as part of the environment. Thus, the word *talawath*, or "pollution," elicited a more definite response and one more in accord with our research project.

Environment and Pollution

We tried to find out what people understood by the environment, since they saw that there was at least some kind of problem with it. Of the total respondents, 11.3% said they did not know. Of the others,[55] 39.1% gave a broad answer, "everything around us," while 42.9% said "the place, the district," which we have taken as similar answers. Another 14.8% said "people, society," focusing on the human dimension.

Overall, 5.1% of these respondents cited some kind of threat or pollution in their definition of "environment," and another 4.4% mentioned cleanliness. Those who mentioned a threat or danger were 12% of the people answering in Kafr al-Elow, and 0.2% in Abkhas, with Dar al-Salam (4.8%) and Sayyida Zeinab (1.9%) in between. The genders were about equally balanced on this answer (5.2% for women, 5.1% for men).

Relatively few people mentioned pollution as a threat or danger under the heading of environment. However, more people were willing to define pollution (96.7%) than environment (88.7%). Overwhelmingly, the answer to the question of the definition of pollution was garbage (45.2% of people answering overall, ranging to a high of 52.6% in Dar al-Salam; women were slightly more likely to give this answer than men, 47.3% to 43.3%). Much less frequent was the second most common answer, air pollution (17.4%

overall, with a high of 31.1% in Sayyida Zeinab, but note that 23.4% of people in Kafr al-Elow cited "cement," which affects people through its pollution of the air). The next most frequent definition of pollution was "anything endangering health" (i.e., a health hazard; 12.1%). Then followed sewage (9%), "quality is not good for people" (8.4%), mosquitoes (8%), dust (7.6%), diseases (6.8%), cement (6.3%, all from Kafr al-Elow), flies (6%), noise (4.3%), water (4.3%), the canal (3.4%), and minor answers. Sewage and garbage were most cited as part of the definition in Dar al-Salam, flies and mosquitoes in Abkhas, noise and dust in Sayyida Zeinab, and cement dust and canal problems in Kafr al-Elow. Gender differences were not striking. Overall, we can sum up most of what people mean by pollution by glossing it as "dirt." We can assume that these answers are not so much a definition as a reflection of a felt problem.

Most people identified polluted food as unclean food, uncovered and vulnerable to flies, or simply spoiled.[56] About 70% gave this answer. Those who saw food as polluted by microbes were 6.9%, by chemical spraying were 6.9%, by cement dust were 7.0%, while smaller numbers saw food as polluted by irrigating with sewage water, by irrigating with factory wastes, by excessive preservatives, and by nuclear radiation.

Similarly, polluted water is seen as simply turbid (i.e., visibly containing sediment) by 32.3% of the respondents.[57] Other common definitions of "polluted water" were that water is polluted when it is affected by sewage water or canal water or by garbage, when it is stored or uncovered, when it contains microbes, when it is mixed with factory wastes, when it contains chemicals, or when it comes from a polluted environment. A few used a smell to identify polluted water. It can be concluded from this that most people judge water to be polluted either by its origin or its appearance.

Nature

In our 1997 survey, we asked people to define "nature," keeping in mind the importance of nature in American environmental concepts (Kempton et al. 1995). People's understandings of the meaning of "nature" included a number of ideas. Overall, 23.1% chose an answer that referred to air or weather: "nice weather or atmosphere; sun, clean air, pure air," while 18.9% referred to greenery, countryside, fields. Another 11.6% invoked the idea that nature was everything that God created and had not been changed by human beings. The people of Abkhas were less likely to give the first two answers, and more likely to give the third one.

Other answers with more than 1% of the total sample were: "nice areas" (6.6%), "a healthy environment, good living with no pollution" (3.1%), "life in general" (3%), "beauty and tranquillity" (2.7%), "religious concerns" (2.6%), "each person has its own nature, good or bad" (2.3%), "every living thing" (2.3%), "water, air" (2%), and "safety, good people, cooperation" (1.2%). Some answers were pretty distant from our meaning of the term, for instance, those who mentioned family or work concerns. Altogether there were 32 unclassifiable responses that were not answers to the question, 167 who said they did not know, and 89 answers that were simply missing,[58] for a total of 12.5% of all answers.

In general we can see that for most people nature is clean air, sunshine, and greenery, a physical description. A minority of just under 15% stressed the identity between God's creation and nature or otherwise gave the concept a religious turn. There was no very obvious difference by gender.

Environmental Concern

We were interested in determining the level of concern with environmental issues. Our first question (1995, see Table 8) about the environment per se was: "Do you consider yourself concerned about the environment?" Overall 50.4% said they were very concerned, 32.3% said they were partly concerned, and 17.3% said they were not concerned. However, there were clear differences according to the four zones. The highest level of concern is in Kafr al-Elow, and the lowest in Abkhas, with Sayyida Zeinab and Dar al-Salam in between, which is a clear reflection of the circumstances of those localities.

Table 8: Responses to the question: "Do you consider yourself concerned about the environment?"

		Locality				*Gender*	
Level of concern	*All*	*KE*	*DS*	*SZ*	*AB*	*F*	*M*
				(%)			
Very	50.4	59.4	53.5	50.4	34.7	47.6	53.0
Partly	32.3	24.1	29.8	35.6	41.7	33.1	31.6
Not	17.3	16.5	16.7	14.0	23.6	19.3	15.4

KE=Kafr al-Elow, DS=Dar al-Salam, SZ=Sayyida Zeinab, AB=Abkhas.

F=Female, M=Male.

Source: 1995 survey. Percentages are of the whole sample (N_{1995}=2,266).

Men were slightly more likely to identify themselves as "very concerned" (53% to 47.6% for women; see Table 9). There was no pattern by age group, but those who identified themselves as "very concerned" rose steadily with education from 35.1% for illiterates to 64.4% for those with a university degree.

This measure of concern should be compared to the answers to the question: "Do you think that those around you are concerned about the environment?" People are nearly twice as likely to see themselves rather than those around them as very concerned (50.4% to 28%), with roughly the same result in each locality and by gender. This suggests that people must not discuss these issues very much, hence they know their own feelings much better than those of others (but see below). But this result is also consistent with a general tendency not to trust other people and to see them as motivated only by individual interest.

Table 9: Respondents' perceptions of those who are "very concerned" about the environment

		Locality				*Gender*	
	All	*KE*	*DS*	*SZ*	*AB*	*F*	*M*
				(%)			
Myself	50.4	59.4	53.5	50.4	34.7	47.6	53.0
Those around me	28.0	32.6	37.4	20.8	19.6	29.0	27.9

KE=Kafr al-Elow, DS=Dar al-Salam, SZ=Sayyida Zeinab, AB=Abkhas.

F=Female, M=Male.

Source: 1995 survey. Percentages are of the whole sample (N_{1995}=2,266).

Overwhelmingly (95% overall) people thought that environmental pollution was a major problem in their village or district. The affirmative answers ranged from unanimity in Kafr al-Elow to a low of 85.3% in Abkhas, with Dar al-Salam at 96.8% and Sayyida Zeinab at 95.8%. These figures confirm the self-perception of Kafr al-Elow, as a heavily polluted site, a perception that is also shared by the media. Conversely, although concern in Abkhas is still high on an absolute scale, the relatively low level of concern there is consistent with the idea (here clearly shared by rural people themselves) that rural areas are unpolluted.

Optimism

Despite everything, most people describe themselves as optimists. In 1995 we asked respondents whether they would describe themselves as an optimist or a pessimist with regard to the environment, with two further intermediate options of "neither" or "sometimes optimist." Overall, nearly four fifths of the respondents (79.8%) described themselves as optimists in 1995, and this rose to 91.1% in 1997 (see Table 28). In the next chapter, Table 40 gives the breakdown of this figure by locality, gender, age, and education, and compares it with other results.

The most discordant note in this symphony of optimism comes from the residents of Kafr al-Elow, who were about 10% below the general figure. Indeed, 14.8% of the Kafr al-Elow respondents described themselves as pessimists. The lowest figure for pessimists was in Abkhas (1.7%, although 18.3% described themselves as "neither" or "in between").

Another measure of overall concern for environmental issues was whether the respondent discussed pollution with friends or relatives. Somewhat less than half of the overall sample (38.1%) said that they had discussed pollution. This figure rose with education, so that 58% of the university graduates but only 26% of the illiterates said that they did. The respondents in Dar al-Salam (38.6%) and Sayyida Zeinab (38.3%) answered this question affirmatively at about the same rate as the overall sample, while those in Kafr al-Elow were much more likely to discuss pollution (56.1%) and those in Abkhas much less likely (14.3%). Men were slightly higher than the overall figure (40.7%) while women were slightly lower (35.3%). On the whole, the most obvious differences between respondents are shown by the increase in positive answers to this question with education, and the contrast between the two localities of Kafr al-Elow and Abkhas.

Ranking of Concerns

From general definitions, we moved on to inquire into the relative importance of different problems. We asked several questions to ascertain the relative priorities of various environmental pollution problems in our four localities. They were asked in somewhat different ways, and the answers are not totally consistent, although there is no doubt about the larger picture.

Overall, the area of greatest concern to people is the disposal of garbage and sewage. The next most worrisome issue is air pollution, followed by water pollution and noise. The figures for air pollution are certainly increased by the high level of concern in Kafr al-Elow about airborne cement dust. The concern about sewage problems is highest in the two inner city areas, Sayyida Zeinab and Dar al-Salam, both of which were going through sewage improvements at the time of the study. A surprising finding for us is the frequency with which mosquitoes, flies, and insects in general were mentioned. By and large people recognize that insects result from other problems, notably the nondisposal of household wastes, but they note these pests as a separate problem.[59]

Since most people (95%) accepted that their own neighborhood had such environmental problems, we asked respondents to identify the problems their neighborhood had. Table 10 gives the breakdown of their answers in 1995, and the 1997 answers are discussed below.

These answers were freely given, i.e., not according to a preestablished list. The table lists the major answers, but it was not possible to combine similar answers (such as polluted air and cement dust) because respondents gave multiple answers and might have cited both. Other answers included: unpaved roads, car exhaust and too many vehicles, overcrowdedness, insecticide, and social issues like cultural pollution and illiteracy. A small number of respondents provided no answer: n=75, or 3.3%, of the whole sample. The figures given are thus percentages of the 2,191 who did give an answer.

Table 10: Perceptions of neighborhood environmental problems

Responses to the question:
"What serious problem does your neighborhood have?"

	Locality				
	KE	*DS*	*SZ*	*AB*	*All*
Answer			*(%)*		
Garbage	**54.1**	21.7	**64.2**	34.8	**44.3**
Sewage overflow	—	**38.3**	35.8	0.2	19.9
Noise	0.2	7.5	38.9	2.9	13.0
Polluted air	3.0	19.3	17.8	4.3	11.6
Mosquitoes	2.7	9.2	7.2	33.8	11.5
Cement dust	39.1	—	—	—	10.7
Cesspits	0.2	31.2	—	1.4	8.7
Garbage in canal	6.2	0.5	—	**36.0**	8.6
Lack of sewage system	0.7	20.0	—	14.1	8.3
Insects	0.7	2.5	6.2	25.7	7.4

Multiple answers possible so totals exceed 100%.

KE=Kafr al-Elow, DS=Dar al-Salam, SZ=Sayyida Zeinab, AB=Abkhas.

Highest choice in each locality in bold.

Source: 1995 survey. Percentages are of those who gave an answer (n=2,191).

In addition to these, some items were singled out in particular localities. In Dar al-Salam polluted water was mentioned by 16.3% and water mixed with sewage by 10%, and in Sayyida Zeinab car exhaust was cited by 11.1%.

We can infer from these figures that garbage and sewage (dirty streets) are what most people think of when asked about pollution.[60] This is true for all four sites, with some variation. The inhabitants of the two sites with open canals, Kafr al-Elow and Abkhas, had a certain level of concern for the quality of water in the canals. However, there was relatively little concern in Kafr al-Elow and Abkhas for the quality of water from other sources. The figures for air pollution are surprisingly low, especially for Kafr al-Elow where in fact in most conversations on pollution people complain about the cement dust from the Helwan Portland Cement Co.[61] The low figure in Kafr al-Elow for "insects and rats/rodents" is also surprising. In interviews, people in Kafr al-Elow in fact often complain about the mosquitoes, which they attribute to the canal and the open-air garbage dumps.

Garbage was cited as the biggest environmental problem in Sayyida Zeinab, just ahead of air pollution, while sewage was cited as the biggest problem in Dar al-Salam, ahead of air pollution. It is consistent with what we know of Sayyida Zeinab and Dar al-Salam that garbage is more of a concern in the former, and sewage in the latter. The people of Sayyida Zeinab are by far the most bothered by noise. Overall, Kafr al-Elow is the locality with the highest level of concern about environmental degradation.

We asked a similar question in 1997: Is there pollution in your neighborhood? The answers essentially parallel those from 1995. In this round, almost two thirds of the respondents (63.1%) said that they thought there was pollution in their neighborhood, about equally men and women (61.5% to 64.7% respectively). The feeling was practically unanimous in Kafr al-Elow (98.8%), a majority in Sayyida Zeinab (66.8%) and in Dar al-Salam (56.9%), but was less than a quarter in Abkhas (23.5%). Almost all those who identified a source of pollution thought that it led to increased medical expenses (96.4%). The question of identifying that source allowed for multiple answers, of which there were 2,543. The most commonly given answers to the question "What is the most serious form of pollution in your neighborhood?" (with no marked difference by gender) were:

1. Garbage, water, and the smell of garbage (25.7%, in all four localities).
2. Cement dust (21.9%, but practically all from Kafr al-Elow).
3. Canals, swamps, and *mesqa*s (irrigation canals) (12.4% in Kafr al-Elow and Abkhas).
4. Dust and unpaved roads (9.1%, especially in Dar al-Salam).
5. Sewage and evacuation wagons (7.9%, mostly in Sayyida Zeinab).

6. Smoke and car exhaust (7%, especially in Sayyida Zeinab and Dar al-Salam).
7. Factories of all types (5.2%, predominantly in Sayyida Zeinab and Dar al-Salam) and factory noise (3%, especially in Dar al-Salam and Sayyida Zeinab).

Table 11: Perceptions of environmental threats

Responses to the question:
"What do you consider the most serious environmental problem?"*

	Locality				
	KE	*DS*	*SZ*	*AB*	*All*
Response			*(%)*		
Most frequently cited					
Air	**77.2**	25.4	29.5	13.0	**37.6**
Water	8.5	**39.9**	15.6	**33.8**	23.9
Garbage	6.7	13.3	**34.1**	26.0	19.7
Intermediate frequency					
Noise	0.2	4.9	10.0	0.9	4.1
Population increase	0.2	9.4	2.8	1.3	3.5
Misuse of fertilizers	0.3	1.0	0.7	11.7	3.0
Least frequent					
Disposal of hazardous waste	1.5	1.0	2.0	2.3	1.7
Ozone layer	1.8	2.2	1.0	0.6	1.5
Nuclear radiation	0.5	0.5	2.0	0.6	0.9
Exhaustion of resources	0.8	0.8	0.5	0.9	0.8
Global warming	0.2	0.2	—	0.4	0.1

* People were offered a list of eleven possible environmental problems, and asked to rank them. We have constructed this table by considering only the first choices. We explored various other ways of utilizing this data, and concluded that this was the most straightforward and did not distort the rest of the data. Since people often did not complete the list it is impossible to construct an overall numerical score.

KE=Kafr al-Elow, DS=Dar al-Salam, SZ=Sayyida Zeinab, AB=Abkhas.

Highest score in each column in bold.

Totals are less than 100% because of failures to answer, most common in Abkhas.

Source: 1995 survey. Percentages are of the whole sample (N_{1995}=2,266).

In 1995, we also asked our respondents to identify what they considered to be the most serious form of pollution in general as opposed to in their particular neighborhood. In Table 11, we have grouped the answers into three categories by frequency: a top three, a middle three, and a bottom group. It is interesting to note that, although the figures are dominated by the major forms of pollution, all the minor ones found a few individuals in each locality who gave them top priority (except for global warming in Sayyida Zeinab). Overall these minorities were returned by 5% of the sample, evenly distributed in the four localities. Most of these reflect a more global outlook, nourished (we assume) by the media rather than an immediate practical approach or a reaction to local circumstances.

If the overall figures place air pollution in first place, this is largely because of Kafr al-Elow, where 77.2% of the respondents listed it first. In the other localities, air pollution was in second or third place. Garbage was of greatest concern in Sayyida Zeinab (34.1% put it first) and in Abkhas (26% put it first), and water in Dar al-Salam (39.9% first) and Abkhas (33.8% first). In all locations women were slightly more likely to put air pollution in first place (40% to 35.4%), and men more likely to put water pollution in first place (26.5% to 21.1%). Likewise, garbage was cited by 21.1% of the women and 18.3% of the men overall.

There is a slight tendency for concern for water pollution and garbage to increase with age, and for concern for air pollution to rise until age 45, then decrease. There is also a slight tendency for the unschooled to be more concerned about air pollution and garbage, and for the schooled to be more concerned about water pollution. None of these tendencies is very strong.

We were also interested in how respondents ranked "the environment" compared to other issues, in order to situate environmental concerns within a wider framework. In this question, respondents were given seven common issues and asked to rank them in order of importance to themselves. Poverty and inflation (money) and pollution ranked first and second among both genders, and everywhere except in Abkhas, where family affairs came before pollution. Given the focus of the questionnaire on the environment, those who ranked the environment high were probably more numerous than they might otherwise have been. Counting only first choices puts pollution in front. Weighting the score for all choices puts money in front.

Generally the ranking of concerns is:

1. Poverty and Inflation
2. Pollution
3. Family Affairs
4. Population Increase

5. Death
6. Wars
7. Crime

Table 12, based on the same question, shows those problems that were of most concern most frequently.[63] (The ranking is a little different than if all choices are weighted, as in the list above.) The overall ranking is mirrored by two of the four localities (Sayyida Zeinab and Dar al-Salam), while Kafr al-Elow and Abkhas rank poverty and inflation before pollution effects as a serious problem. This probably reflects the overall economic situation of the localities. As a sidelight it is interesting that relatively the most worried group about crime is in Abkhas village.

Table 12: Ranking of problems by locality

	Locality				
	KE	*DS*	*SZ*	*AB*	*All*
Most pressing problem			*(%)*		
Pollution	33.6	40.3	40.6	24.0	35.2
Poverty & inflation	37.6	24.0	28.8	39.1	32.0
Family problems	7.5	14.5	10.6	15.1	11.7
Population increase	7.0	11.3	9.4	9.8	9.4
Wars	7.7	3.5	6.6	2.6	5.2
Crime	1.7	1.8	2.0	3.2	2.1
Death by accident	0.8	2.3	1.2	3.0	1.8

KE=Kafr al-Elow, DS=Dar al-Salam, SZ=Sayyida Zeinab, AB=Abkhas.
Highest score in each column in bold.
Totals are less than 100% because of failures to answer.
Source: 1995 survey. Percentages are of the whole sample (N_{1995}=2,266).

It is interesting to compare these results with the concerns of the Egyptian government as expressed in the NEAP of 1992 and the documents prepared for the Earth Summit in Rio de Janeiro in 1993. Although the concerns cited in NEAP were deliberately not prioritized, one can infer from the listing given and from the priority list prepared for Rio that the protection of natural resources (green space, freshwater, marine resources, and soils) ranked first. This certainly includes the need to guarantee Egypt's fresh water supply from the Nile. Then came the necessity to protect against air and water pollution, and the need to deal with solid and liquid waste. In the Rio list, "supporting environmental awareness and people's participation" was cited last, after "developing environmental management instruments" (Gomaa 1997:38).

From another point of view our results can be compared to the ranking in Ibrahim (1995:74): inflation and low incomes, then housing, terrorism, unemployment, too many people, declining morality, and corruption are listed there. It is not impossible that pollution scores so high in our study, while it is absent in Ibrahim's, because it was clear to the respondent that pollution was what we were interested in. In our ranking, money and pollution scored very close to each other, so that sometimes one and sometimes the other ranked first, depending on the tabulation method.

Causes of Pollution

We pursued this question of ranking by asking our respondents to identify the major sources of air, water, and noise pollution. The answers are given in the following tables, broken down by locality (Tables 13 through 15) and by gender (Tables 16 through 18). The questionnaire included a number of the most obvious alternatives, including all the ones listed here with the exception of "workshops," which was amalgamated with "factories and workshops" under noise pollution and represents about half that figure. Also, "no air/water/noise pollution" was not a listed option but was sometimes volunteered, especially for noise in Abkhas.

Table 13: What is the major source of air pollution?

	Locality				
	KE	*DS*	*SZ*	*AB*	*All*
Source			*(%)*		
Factories	**82.4**	30.0	30.6	7.4	**39.4**
Vehicles	2.7	**46.6**	**50.4**	28.3	32.2
Dust	14.1	16.2	14.5	**50.0**	22.2
Total	99.2	92.8	95.5	85.7	93.8

KE=Kafr al-Elow, DS=Dar al-Salam, SZ=Sayyida Zeinab, AB=Abkhas.

Highest choice in each locality in bold.

Totals are less than 100% because of failures to answer.

Source: 1995 survey. Percentages are of the whole sample (N_{1995}=2,266).

Table 14: What is the major source of water pollution?*

	Locality				
	KE	*DS*	*SZ*	*AB*	*All*
Source			*(%)*		
Dumping in the Nile	**69.0**	**81.4**	**67.9**	**68.8**	**72.0**
Factory waste	21.6	9.6	18.8	5.3	14.3
Washing in the canal	1.7	3.2	9.7	22.8	8.6
Dust	5.7	1.5	3.2	1.3	3.0
Total	98.0	95.7	99.6	98.2	97.9

* understood in terms of Nile/canal pollution
KE=Kafr al-Elow, DS=Dar al-Salam, SZ=Sayyida Zeinab, AB=Abkhas.
Highest score in each column in bold.
Totals are less than 100% because of failures to answer.
Source: 1995 survey. Percentages are of the whole sample (N_{1995}=2,266).

Table 15: What is the major source of noise pollution?

	Locality				
	KE	*DS*	*SZ*	*AB*	*All*
Source			*(%)*		
Overcrowding	23.0	**41.6**	24.5	20.6	**27.8**
Vehicles (all types)	**27.8**	26.3	**32.3**	21.3	27.3
Microphones, radios, TV	22.8	21.7	26.1	9.6	20.7
Factories & workshops	16.9	6.8	16.1	0.9	10.7
No noise pollution at all	8.3	0.7	0.2	**46.8**	12.1
Total	98.8	97.1	99.2	99.2	98.2

KE=Kafr al-Elow, DS=Dar al-Salam, SZ=Sayyida Zeinab, AB=Abkhas.
Highest score in each column in bold.
Totals are less than 100% because of failures to answer.
Source: 1995 survey. Percentages are of the whole sample (N_{1995}=2,266).

Table 16: What is the major source of air pollution?

	Gender		
	F	*M*	*Both*
Source		*(%)*	
Factories	37.4	41.2	39.4
Vehicles	34.3	30.1	32.2
Dust	22.5	21.9	22.2
Total	94.2	93.2	93.8

F=Female, M=Male.

Totals are less than 100% because of failures to answer.

Source: 1995 survey. Percentages are of the whole sample (N_{1995}=2,266).

Table 17: What is the major source of water pollution?

	Gender		
	F	*M*	*Both*
Source		*(%)*	
Dumping	**76.1**	68.2	72.0
Factory waste	10.7	**17.9**	14.3
Washing in the canal	7.6	9.5	8.6
Dust	3.3	2.8	3.0
Total	97.7	98.4	97.9

F=Female, M=Male.

Highest choice by row (female vs. male) in bold where the difference is greater than 5 percentage points.

Totals are less than 100% because of failures to answer.

Source: 1995 survey. Percentages are of the whole sample (N_{1995}=2,266).

Table 18: What is the major source of noise pollution?

	Gender		
	F	*M*	*Both*
Source		*(%)*	
Overcrowding	26.6	28.8	27.8
Vehicles (all types)	**30.3**	24.3	27.3
Microphones, radios, TV	18.5	22.7	20.7
Factories and workshops	10.6	10.8	10.7
No noise pollution	12.3	12.0	12.1
Total	98.3	98.6	98.6

F=Female, M=Male.

Highest choice by row (female vs. male) in bold where the difference is greater than 5 percentage points.

Totals are less than 100% because of failures to answer.

Source: 1995 survey. Percentages are of the whole sample (N_{1995}=2,266).

There are some differences between the localities. Kafr al-Elow respondents were most likely to identify factories as the major source of air pollution, while Dar al-Salam and Sayyida Zeinab chose vehicles, and Abkhas dust. While respondents from all four localities pinpointed people dumping material in the canals and the Nile as the major source of water pollution, Kafr al-Elow and Sayyida Zeinab people were more likely to think that factories were a major source of water pollution than people from the other two localities. Not surprisingly, people washing clothes and animals in the canal was the water pollution source most often cited by people from the village of Abkhas. Nearly half the people in Abkhas, and a few elsewhere, said there was no noise pollution at all.[64] For the urban sites, noise pollution is essentially linked to congested urban living conditions. Overcrowdedness per se ranked first in Dar al-Salam (and second among those who gave an answer in Kafr al-Elow and Abkhas), while vehicles ranked first in Kafr al-Elow and Sayyida Zeinab (and first among those who gave an answer in Abkhas). Factories and workshops were ranked first by 16.9% of the respondents in Kafr al-Elow and 16.1% in Sayyida Zeinab (these areas also ranked vehicles as the major source of noise).

There were some minor differences between the genders in the answers to the three questions about the source of air, water, and noise pollution. Women were somewhat more likely than men to think that vehicles were the main source of air pollution (34.3% to 30.1%), and somewhat less likely to think that factories were the cause (37.4% to 41.2%); the answers for dust were nearly equal (22.5% for women and 21.9% for men). Women were more likely than men to identify dumping in the canals as the main source of water pollution though for both it was by far the most common choice (76.1% to 68.2%). Women were more likely to identify vehicles as the major source of noise pollution (30.3% to 24.3%) and somewhat less likely to choose the other possibilities.

Respondents with higher levels of education were more likely to identify vehicles and factories as the major sources of air pollution, while the less educated were somewhat more likely to choose dust. However, "factories" was the top source identified by all the educated groups except graduates of secondary schools. The more educated were more likely to identify factories as the major source of water pollution (23% of university graduates as compared to 9.9% of the illiterate), though like everyone else all groups stressed dumping in the canals. The more educated were more likely to think there was noise pollution, but otherwise the different educational strata followed the general pattern as given in Table 15. Factories were not singled out here as the problem; primary school graduates were the most likely to think that factories and workshops were the major source of noise pollution (15.5% compared to 13.5% for secondary school graduates and 13.4% for technical school graduates). Those who were bothered by microphones, radios, televisions, etc., were a steady ratio of about 20% for all gender, education, age, and locality groups.

There was relatively little difference by age. One can note only a slight tendency, as age increases, to think that vehicles are more of a problem and factories less of one as far as air pollution is concerned. The older groups also worry slightly less about factories as a source of noise pollution, and slightly more about overcrowdedness.

The overall impression from these results is that many respondents find other people are the problem—dumping waste in canals, driving their vehicles, using their microphones, simply being there. Activity in factories and workshops is accepted as a source of air, water, and noise pollution, but not a major one. Impersonal dust is also a major source of air pollution.

Blame and Responsibility

On the question of who or what should be blamed or held responsible for pollution, the largest number (72.1%) thought that people in general were at fault.[65] Another sizable minority (35.5%) blames factories, while 15.1% blame cars and transportation, and 9.4% blame the misuse of fertilizers and pesticides. Smaller numbers blamed smoking (3.8%), nuclear experiments (1.4%) and wars (1.4%). Those who blamed the government were 7.8%.[66] There were only 0.7% "don't know" answers.

These figures showed some variation by locality. Thus, in Kafr al-Elow there were predictably more who blamed factories than blamed people in general (70.4% to 49.6%). Next most likely to blame factories was Sayyida Zeinab, where 34.3% of the respondents included this answer, compared with 84% who cited people in general. Cars and vehicles were seen as the major culprit by 23% of those in Sayyida Zeinab, and 22.3% of those in Abkhas, while they were singled out by 12.3% of those in Dar al-Salam and 4.3% in Kafr al-Elow. Those who blamed the government were most numerous in Dar al-Salam (13.5%), followed by Kafr al-Elow (8.7%), then Sayyida Zeinab (5.2%) and Abkhas (2.8%). The misuse of fertilizers and pesticides was predominantly cited in Abkhas, where it was mentioned by 29.1% of the respondents. It was also cited by 5.8% in Sayyida Zeinab, 3.7% in Kafr al-Elow, and 3% in Dar al-Salam. The largest group of those who blamed smoking were in Sayyida Zeinab (6.3%), followed by Abkhas (4.7%), Kafr al-Elow (2.5%), and Dar al-Salam (1.9%).

There was relatively little variation by gender or age regarding who or what the respondents thought was responsible for pollution. The main gender difference is that men simply gave more answers to this question where multiple answers were possible. Among the possibilities they were more likely than women to cite were traffic and the misuse of fertilizer. The higher the educational level, the more respondents blame people in general, factories, and traffic (of the post-secondary respondents, 82.9% blame people in general, 38.9% blame factories, and 19.6% blame vehicles). There is little pattern with respect to the role of government and the misuse of fertilizers.

We asked a slightly different question in 1997, but the general answer was similar. Most people agreed with the statement that people are the cause of environmental pollution, and cited poor handling of garbage as the reason (70.9%). Others who thought that people were to blame pointed out that the factory owners were people (2.9%), or noted that people start a project without thinking of its impact on others (0.6%). On the other hand, some disagreed, and pointed out that the problem was services (10%), factories (6%), or the government (3.2%). All these answers came mostly from Kafr al-Elow. There was no striking difference between the genders in these answers.[67]

Table 19: Perceived environmental change over the last five years

Over the last five years, do you think the environment in your community has become better or worse?

		Locality				*Gender*	
	All	*KE*	*DS*	*SZ*	*AB*	*F*	*M*
Answer				*(%)*			
Better	**29.4**	22.0	**38.9**	24.6	**33.2**	26.0	**32.7**
Worse	27.9	**37.8**	25.6	**29.8**	15.7	**30.4**	25.5
No change	42.6	39.9	35.5	45.6	51.1	43.5	41.8
Total	100.0	99.7	100.0	100.0	100.0	99.9	100.0

KE=Kafr al-Elow, DS=Dar al-Salam, SZ=Sayyida Zeinab, AB=Abkhas.
F=Female, M=Male.
Highest figure between "better" and "worse" in bold.
Totals are less than 100% because two individuals said they did not know.
Source: 1995 survey. Percentages are of the whole sample (N_{1995}=2,266).

Table 20: Perceived changes in drinking water over the past five years

How has the quality of drinking water in your neighborhood changed?

		Locality				*Gender*	
	All	*KE*	*DS*	*SZ*	*AB*	*F*	*M*
Answer				*(%)*			
Better	**23.3**	**31.6**	21.7	**22.8**	**15.3**	**26.0**	**20.7**
Worse	15.5	7.8	**23.4**	16.1	14.5	15.8	15.2
The same	61.2	60.6	54.9	61.1	70.2	58.2	64.1
Total	100.0	100.0	100.0	100.0	100.0	100.0	100.0

KE=Kafr al-Elow, DS=Dar al-Salam, SZ=Sayyida Zeinab, AB=Abkhas.
F=Female, M=Male.
Highest figure between "better" and "worse" in bold.
Source: 1995 survey. Percentages are of the whole sample (N_{1995}=2,266).

Village leaders from Abkhas

Street in the village of Abkhas

Women washing dishes in the main canal in Abkhas

Morning rush hour at the metro station in Dar al-Salam

A tree grows in Dar al-Salam after sewage construction

Government sewage system under construction in Dar al-Salam

Residents of Dar al-Salam cope with a community-built sewage line

Fayoum Street, Dar al-Salam

Street vendors, Dar al-Salam

Garbage smolders next to an outdoor kitchen in Dar al-Salam

Researchers relax in a café in Dar al-Salam: Necla Tschirgi, Nicholas S. Hopkins, Sohair R. Mehanna, Said Samir, and others

Women pump water on a garbage heap in Kafr al-Elow

The cement factory looms over Kafr al-Elow

Smoke from a workshop in Kafr al-Elow

The Khashab Canal in central Kafr al-Elow. The awning covers a vegetable vendor's stall

A street in Ma'adi

A street in Sayyida Zeinab

A donkey-drawn garbage collection cart in Sayyida Zeinab

The swings at the annual *mulid* in Sayyida Zeinab

Contrasts

Time Trends

We asked whether the environment of the community had gotten better or worse over the previous 5 years (see Table 19). The most common answer, expectedly, was that it had stayed the same—42.6%, compared with 29.4% who thought it had improved and 27.9% who thought it had deteriorated. Interestingly enough, Kafr al-Elow and Sayyida Zeinab residents were more pessimistic, more likely to think that their community's environment was getting worse, while Dar al-Salam and Abkhas residents thought it was improving. Women were slightly more likely to think that the environment had become worse, and men to think that it had improved. The optimists outnumbered the pessimists in all age groups except the one between 26 and 35, and in all educational levels except the illiterates and those educated to primary school level, though the spread was slight.

People are more likely to think that the environment has changed, for better or for worse, than they are to think that the quality of drinking water has changed. Among those who think that there has been a change, there is more concern about the deterioration of the environment in general than about the quality of drinking water (27.9% to 15.5%). Among the localities, people in Kafr al-Elow are quite optimistic about the quality of their drinking water. People in Dar al-Salam, who are pessimistic about drinking water, are optimistic about the environment in general; the optimists are perhaps thinking of the construction of the sewage system. There were slight tendencies for women to be more optimistic about the quality of drinking water than men (the figure for those who thought the quality was worsening is the same, but there are more among women who think it is improving, and more among men who think it is staying the same), and for pessimism to increase with education (those at the postsecondary level were the most likely to think that drinking water quality was worsening and least likely to think it was improving, but by small margins).

Seasonal

On the whole, people find the environmental conditions in winter are less stressful than those in summer. We asked respondents to compare the state of the environment in their neighborhood in the winter and in the summer. Overall, 35.3% saw no difference, and this figure was remarkably stable over the localities, genders, ages, and levels of education (the highest for any of these categories was 36.9%; the lowest was 33.1%). Of those who sensed a difference, the general feeling was that the environmental conditions were worse in the summer, generally by a ratio of nearly 4:1. Overall 50.8% said environmental conditions were worse in summer and 13.9% said they were worse in winter. The people of Sayyida Zeinab were most likely to feel that summer was difficult, since 56.9% said that conditions were poorer then. Men and women had nearly the same score here. Those with post-secondary education were somewhat more likely to find summer worse than others (57.4%, with other educational levels falling between 46% and 51.9%).

Work More Dangerous than Home

Of the 1,041 respondents in 1997 who were working and thus answered the question on whether the workplace was more dangerous than the home, 706 (67.8%) found that it was. There was considerable variation by locality. Only in Kafr al-Elow did a minority of people agree that work was more dangerous than home (41.8%). In the other two urban sites, 65.5% felt this in Dar al-Salam and 68% in Sayyida Zeinab, while almost everyone (93.5%) in Abkhas agreed. The reason for this feeling in Kafr al-Elow is pretty obvious: the cement plant near the residences makes them hazardous. In the other localities people focused on the nature of work and the polluted air found in the workplace to reach their conclusion. Interestingly, some respondents mentioned exposure to smokers as one hazard of work.

Women who answered the question were evenly divided over the risks of home and work (49.6%), while 70.6% of men found work more hazardous. Among those who found work more dangerous, men tended to stress the nature of work while women stressed the polluted environment (air). Women, who are more likely to spend time in the home, are relatively speaking more likely to find that the conditions there are hazardous.

Relative Effects

We asked our 1997 respondents to compare the effects of pollution on various paired categories. Some of the results are given in Table 21. This table also shows the rank ordering of the five propositions. It is noteworthy that in three of the four localities, except Abkhas, and for both genders, the rank ordering is the same.

There was almost total consensus on the idea that the poor are affected more than the rich, and children more than teenagers. A very large majority favored the idea that elders were more affected than youth, and that workers were affected more than nonworkers. The latter idea is consistent with the result that people generally find the workplace more threatening than the home.

Opinion on the relative effect on the genders was evenly split, with a narrow minority in support of the idea that women were more affected. Even

Table 21: Perceived relative effect of pollution on different categories

Pollution effect	*All*	*Locality*				*Gender*	
		KE	*DS*	*SZ*	*AB*	*M*	*F*
Greater : lesser				*(%)*			
Poor : rich	98.3	99.0	98.8	98.5	96.4	97.9	98.7
Children : teenagers	97.0	95.7	96.8	97.5	98.2	97.4	96.6
Elders : youth	85.9	83.0	88.5	91.0	80.0	87.6	84.1
Workers : nonworkers	83.3	72.9	80.4	90.0	91.1	84.3	82.2
Women : men	52.4	54.1	47.6	42.6	68.0	45.0	60.2

KE=Kafr al-Elow, DS=Dar al-Salam, SZ=Sayyida Zeinab, AB=Abkhas.

F=Female, M=Male.

Source: 1997 survey. Percentages are of the whole sample (N_{1997}=2,307).

though women were more likely to think that women were more affected, and men more likely to think that men were more affected, the figures for each do not approach the clear majorities found in the answers to the other four questions. Thus, 45% of the men thought that women were more affected, and nearly 40% of the women thought that men were more affected. Furthermore, one can see that in two of the four localities the majority favored the idea that men were more affected. Only the relatively high score in Abkhas tipped the balance, so to speak, in favor of the idea that women are more affected.

Rural-Urban and Class Differences

In 1997 we also asked our respondents two questions on their evaluation of the behavior of others. In one question, we asked them whether they agreed with the statement: "People are less cooperative now than they used to be." Table 22 gives the breakdown of the "yes" answers by locality and gender. We can see that overall, nearly nine tenths of the sample agreed with the statement, but that there were some interesting differences by locality and gender. The consensus in agreement with the statement indicates a fairly strong cultural pattern. There is a significant contrast between the two relatively rural localities of Abkhas (a village) and Kafr al-Elow (which retains some village-like ways despite its urbanization) both marked in Table 22 with asterisks, and the two relatively urban ones of Dar al-Salam (new urban) and Sayyida Zeinab (old urban). The percentages of agreement are lowest in the least urban locality (Abkhas), and gradually increase with the degree of urbanization. One could also contrast the "rural" pair with the "urban" pair. This contrast could also reflect class differences, within the broad consensus. Support for this statement is slightly higher among men than women.

Table 22: Direction of change in cooperativeness

Agreement with the statement: "People are less cooperative than they used to be."

	(%)
All	88.6
Locality	
Abkhas*	80.0
Kafr al-Elow*	84.2
Dar al-Salam	92.8
Sayyida Zeinab	96.0
Gender	
Male	90.6
Female	86.5

Source: 1997 survey. Percentages are of the whole sample (N_{1997}=2,307).

Of the minority who felt that people are more cooperative now than in the past, about half (49.4%) attributed their answer to the friendly neighborhood in which they live (more women than men by 54.5% to 42.3%). Others argued that this was true because people are educated now (26.2%; 36.9% men:18.4% women), while a few argued that people interact more now (6.8%; 8.6% women:4.5% men). Other answers accounting for 13.7% of the minority included the idea that this is so because we all suffer from the same problems, or because we are better informed, or because rich people help poor people, or because religion is now more strict and people follow its mandates.

The majority who thought that people are less cooperative nowadays argued that this was so because there is no close contact or love among the people (31.9% of the subset, most common in Abkhas where it was 44.1%), or because people are now only concerned about themselves for economic reasons (24.1%, more common in the three urban sites), or that there is no cooperation among people (17.5%, most common in Abkhas), or that people have become selfish (10%, most common in Kafr al-Elow). There was little gender difference in these answers. All these answers reduce to pointing out the individualism that many people believe marks contemporary Egyptian life. Only one answer was a little bit different: those who attributed the change to the rapid pace of life (6.2%, more in Dar al-Salam and Sayyida Zeinab). At one level, these answers amount to saying that there is less cooperation because people do not cooperate any more, having become more individualistic. Clearly a real sociological analysis of the issue has to probe this tautology further.

Our second question was to ask our respondents how they explained the behavior of people who throw water and garbage in the streets, since many deplored this behavior. Two main answers emerged: first, people have no other choice (55.3% of all respondents) and second, people are illiterate, uncultured, and not aware (31.3% of all respondents). At one level, of course, these answers are a reflection of people's reasons why they themselves might have thrown water and garbage in the streets.[68]

However, there is again an interesting pattern of difference between the localities and the genders. Respondents from Kafr al-Elow and Abkhas were four times more likely to believe people had no choice (93.7% and 92.9%, respectively) than respondents in Dar al-Salam and Sayyida Zeinab (23% and 17.4%, respectively). This is a striking division between these two pairs of localities and the locality differences in the secondmost common answer is also suggestive: In Sayyida Zeinab and Dar al-Salam, more than half attributed the deplorable behavior to illiterate and uncultured individuals (62.8% and 50.4%, respectively). By contrast, only 4.5% in Abkhas and 3.2% in Kafr al-Elow gave this rationale for the behavior. The polarization around this issue, as we have seen, suggests a class division between those who are at the bottom of the social ladder(the two village-like communities) and those who are a rung or two up (the more urban communities). However, the difference in reaction probably reflects more than simply a class contrast. Among other things, it should be remembered that Dar al-Salam and Sayyida Zeinab are better served by garbage collection and sewage than the two other localities. Thus they could afford to blame people rather than circumstances.

Respondents in the two rural localities were much more likely to explain undesirable garbage disposal behavior by citing the absence of alternatives (the relevant figures are in bold in Table 23). They see this antisocial behavior as the result of the lack of alternatives. On the other hand, respondents in the two relatively urban localities see it as due to ignorance and similar notions. In other words, they have a tendency to blame the perpetrators, who are also the victims. The figures for gender do not show significant contrast. Women (57.7% of respondents) are more likely than men (53%) to think that people throw garbage and sewage in the street because there is no choice, while men (34.6%) are more likely than women (27.7%) to attribute this to illiteracy and lack of culture and awareness.

Table 23: Perceived reasons for throwing garbage in the streets

	Respondents answering:	
	No other choice (%)	*Ignorance*[†]
All	55.3	31.3
Locality		
Abkhas	**92.9**	4.5
Kafr al-Elow	**93.7**	3.2
Dar al-Salam	23.0	**50.4**
Sayyida Zeinab	17.4	**62.8**
Gender		
Male	53.0	34.6
Female	57.7	27.7

† Ignorance = illiterate, uncultured, not aware

N.B. Dar al-Salam and Sayyida Zeinab had effective garbage collection and sewerage, in contrast to the other two. Thus, they could afford to blame people rather than circumstances.

Highest choice by locality in bold.

Source: 1997 survey. Percentages are of the whole sample (N_{1997}=2,307).

In these data we can see some elements of a dialectic within our sample whereby the very poor claim that they have no choice in how to behave, while those slightly better off can use that difference to establish a hierarchy of moral virtue. They can use this difference to establish a boundary with themselves on the correct side. Meanwhile, although all four communities have strong majorities that feel that people cooperate less than they used to, this feeling is strongest in the two relatively well-off communities. Thus, all things considered, it is in the poorer pair that people are more likely to think that others are more cooperative and thus that cooperation works.

El-Ramly's material from Ma'adi supports this argument. The explicitly class-based arguments of a small sample of middle- to upper-class women stress the contrast between a clean and aware upper class and a dirty, unaware, and barely correctable lower class in the popular neighborhoods. "The majority of the women interviewed in Ma'adi . . . attribute the responsibility for the deterioration of environmental conditions in Egypt to the 'ignorant' and 'illiterate' Egyptians. Lack of awareness, poverty, overpopulation, and the difficulty of life conditions have also been cited by Ma'adi women as reasons for environmental degradation" (el-Ramly 1996:142).

The Ma'adi women agreed that Egyptians needed to become more aware of environmental issues and problems, but generally thought that it was not they themselves who needed this awareness but those living in popular and low-income areas. They see themselves as "good," living in "high class" (or *raqia*, civilized) neighborhoods, and as "already aware." One woman explained that "environmental conditions could improve in high-income districts such as Ma'adi, Heliopolis, and Mohandiseen, but the same is not true for lower-income districts where people's ignorance, the lack of awareness, and overpopulation only perpetuate pollution problems" (el-Ramly 1996:106). They "do not observe personal cleanliness." They should be educated or informed. As another woman said, those "who live in areas like 'Arab al-Ma'adi and al-Basatin, the illiterate, ignorant, and uncivilized who need to become more aware of environmental problems and pollution," while "the people here in Ma'adi know all that." Consequently, television programs should be aimed at the poor, and social workers should visit them "to give them lectures because they are ignorant." The epistemological model is straightforward: "If people know that something is wrong, they will stop doing it" (p. 127–29). The ultimate solution according to these informants is to raze and rebuild poor neighborhoods (p. 60).

Opinions (1995 and 1997)

In our second survey we repeated a few questions we had already asked in our first survey (see Tables 24 through 28). The general tendency of the answers was to show a slightly more positive, or less concerned, outlook on environmental issues. Respondents overall were a little more optimistic, although even in 1995 nearly 80% of respondents described themselves thus. They were slightly more likely to think that the present generation has a responsibility to pass a clean environment on to the next generation. They were not quite as likely to think that there was an unavoidable conflict between building factories and a clean environment, and they were more willing to have factories even if these were a threat to a clean environment. Those who were "concerned" about the environment in 1997 outnumbered those who were "very concerned" in 1995, although they were less than the total of those who expressed at least a little concern in the first survey. By and large the breakdown by locality follows the patterns described above without surprises. For instance, while the number of optimists rose, the rank order of the four localities did not change. The major shift among these answers is probably the increase in those willing to accept factories despite the threat of pollution from them.

Female respondents in 1997 agreed more readily with each of these five statements. Women are somewhat more likely than men (56.8% to 49.5%) to think that the people around them are concerned about the environment. The number of women who thought that those around them were concerned about the environment appears to have increased faster than the number of men over this two-year period. Women are also more likely than men (48.7% to 38.1%) to think that building new factories would conflict with a clean environment. They are more likely (14.1% to 8.6%) to prefer a clean environment over new factories or (the most popular option) both a clean environment and new factories. They are also more likely (99.1% to 98%) to think that the present generation has a responsibility to pass a clean environment on to the next generation. And they are a little more likely (92.3% to 89.9%) to consider themselves optimists with regard to environmental problems.

Table 24: Others' concern for the environment

Responses to the question: "Are the people around you concerned about the environment?"

		Locality				*Gender*	
	All	*KE*	*DS* (%)	*SZ*	*AB*	*M*	*F*
Level of concern 1995							
Very	28.0	32.6	37.4	20.8	45.7	27.2	28.9
Somewhat	43.2	40.8	39.1	49.3	43.6	—	—
Total	71.2	73.4	76.5	70.1	63.3	—	—
Level of concern 1997							
Yes	53.1	55.9	55.6	53.9	45.7	27.9	48.7

Table 25: Factories conflict with a clean environment

Agreement with the statement: "Building new factories would conflict with having a clean environment"

		Locality				*Gender*	
	All	*KE*	*DS*	*SZ*	*AB*	*M*	*F*
Year			(%)				
1995	54.4	58.2	52.5	55.1	50.9	58.3	50.6
1997	43.3	48.9	42.3	37.6	44.7	55.1	44.9

Table 26: A clean environment preferred to new factories

Agreement with the statement: "I prefer having a clean environment to having new factories"

		Locality				*Gender*	
	All	*KE*	*DS*	*SZ*	*AB*	*M*	*F*
Year			*(%)*				
1995	37.7	40.6	39.2	35.1	35.5	41.9	33.7
1997	11.3	17.0	14.0	9.3	3.6	61.2	38.8

Table 27: The responsibility of the present generation

Agreement with the statement: "The present generation has the responsibility to pass a clean environment on to the next generation"

		Locality			
	All	*KE*	*DS*	*SZ*	*AB*
Year			*(%)*		
1995	95.1	97.7	95.3	94.5	92.6
1997	98.6	98.5	98.2	98.7	99.0

Table 28: Environmental optimism

Agreement with the statement: "I consider myself an optimist with regard to the environment"

		Locality				*Gender*	
	All	*KE*	*DS*	*SZ*	*AB*	*M*	*F*
Year			*(%)*				
1995	79.8	72.2	85.4	81.9	80.0	80.1	79.6
1997	91.1	81.7	95.2	94.3	93.5	92.3	89.9

For tables 24–28,

KE=Kafr al-Elow, DS=Dar al-Salam, SZ=Sayyida Zeinab, AB=Abkhas.

F=Female, M=Male.

Source: 1995 and 1997 surveys. Percentages are of the whole sample (N_{1995}=2,266, N_{1997}=2,307).

Choices

Factories and Environment
Overall a slight majority of the 1995 respondents (54.4%) agreed that new factories would conflict with environmental protection.[69] This feeling was pretty evenly spread among the four localities, though predictably Kafr al-Elow respondents agreed the most (58.2%) and Abkhas the least (50.9%). Women were more likely than men to think that factories would conflict with environmental protection (58.3% to 50.6%). The picture with regard to age and education was less clear.

Despite the opinions on the conflict, fewer were prepared to draw strong conclusions. Of the 41.8% of respondents willing to make a choice between factories and the environment, 90.2% (or 37.7% of the whole sample) said that cleaning up the environment was the higher priority, and 76% of these thought that building new factories would conflict with a clean environment. Women were again more likely to choose environment over factories (41.9% to 33.7%). More in the older group preferred this choice than in the younger group (40.2% to 31.7%), while by and large so did the less well educated (43.9 of the unschooled against 32.9% of the schooled). It should be remembered that most of those who did not opt for environment over factories favored an effort to have both. A majority of 58% said that factories should be built *and* the environment protected.

Our 1997 sample was more or less evenly split on the question of whether building new factories would threaten a clean environment. The figures were pretty consistent across the four localities, though as expected Kafr al-Elow is the highest, about 11 percentage points higher than the lowest, Sayyida Zeinab. We asked the respondents to explain their answers. Those who agreed there would be a threat (43.3% of the total sample) mostly stated that factories would increase pollution, smoke, exhaust, and noise (80% of this subset). Others stressed that factories should be built away from residential areas (6.7%), which is more a recognition of the factories as a nuisance than a general statement of the threat that industrialization poses, while some noted that factories are a source of illness (5.2%). Those who saw no conflict stressed that factories could be sited in distant places in industrial areas (49.9% of answers from this subset). Others noted that factories could use new technologies, which would be less polluting (19.9%), or should be pollution free (5.1%), while yet others argued that factories are necessary to provide jobs for the youth (18.6%, evenly distributed by gender and location). Eight individuals (four in Sayyida Zeinab, two in Kafr al-Elow, one each in Dar al-Salam and Abkhas) noted the importance of increasing production and the GNP. Thus the general feeling about this tradeoff is that it is worth trying to have new factories, but there is a recognition that factories are hazardous for their neighbors, and thus that the technology used in them should be clean or the factories should be located away from population centers. There is some particular hostility to cement factories. Men are more likely than women to see the location of factories in distant spots as a solution to the problem (55.7% to 42.7% among those who accept factories and 9.3% to 4.5% among those who prefer a clean environment); women are more concerned with pollution levels (82.5% to 76.9%) and with jobs for the youth (24.9% to 13.6%).

Hazardous Jobs

Overwhelmingly the 1995 sample said they would not accept a high-paying job in a dangerous environment (96.2%), although in fact a certain number of them may be doing that. Predictably the most willing to do this are in Kafr al-Elow, where only 90.5% said they would not. Those who said they would numbered 57, somewhat less than the 71 who identified themselves as cement factory workers; of course not all jobs in the cement plant are equally dangerous. An even greater proportion said they would not agree to their child working in a dangerous environment (97.7%). The issue in these cases may be to know what people consider dangerous, since there are notoriously many children as well as many adults working in what outsiders consider to be hazardous environments (Farrag 1995).

Religion

Between 92% and 97% of the respondents in the four localities felt that the teachings of religion (i.e., Islam) had some relevance.[70] Over three quarters of the respondents in each of the four localities felt that the teachings of religion relate to the environment to a great extent. The range was remarkably uniform, from 77% in Abkhas to 78.4% in Kafr al-Elow. Males were more likely to give this answer than females (81.8% to 73%), and the schooled were more likely to do so than the unschooled (80.6% to 73.5%). Again, bear in mind that more females number among the unschooled. Age made little difference in the answers. At the other extreme, those who felt that religious teachings were unconnected to environmental issues were most numerous in Dar al-Salam (7.4%) and Kafr al-Elow (6.5%). Pollution was most likely to be discussed in sermons in the mosque in Kafr al-Elow and this was reflected in a higher proportion of respondents opining it was than it was not discussed (36.8% v. 35.1%).

Further interviewing by the resident researchers revealed the reasoning behind these answers. Most people associate pollution with dirt and religion with cleanliness because of the ablutions required before prayers, and so this conclusion is easily reached. More detailed analysis of the ways in which the teachings of religion might relate to conservation or stewardship for the next generation were absent. It became evident in the focus groups that people feel that the role of the mosques (and of organized religion in general) is to provide charity for the poor, and this definition of their role means that the mosques and religious leaders are not expected to take the lead in dealing with pollution in people's lives.

There are of course other discourses in Islam on the environment, notably those that stress that the current generation is the steward of the earth and God's creation for future generations. We found that 62.7% of our second sample strongly agreed with the proposition that "We know that nature is God's creation, so it is wrong to misuse it," while 60.1% strongly agreed that "We are responsible for using our resources such as land and water in the best possible way without being extravagant." The idea of stewardship is present in both propositions.[71] People agree with these ideas, but apparently do not discuss them since they were not offered as comments on other questions.

Focus group participants in all four localities tended to limit the role of religion. Although religious institutions could teach people about the link between religion and the environment, the role of religious teaching would be muted as long as the situation is marked by poverty, lack of community incentives, government laxity, and the lack of resources to support local initiatives. Islam requires cleanliness and hygiene, according to verses from the Quran. Mosques should raise money for the poor and the orphans, and for charitable work in general. Some participants felt that religious leaders were not qualified to play a role outside the religious context.

Knowledge of Laws

Overall, 45.5% of our respondents said they had heard about the new (Law 4 of 1994) environment law, about a year old at the time of the 1995 survey. However, more of them (71.1%) were aware of the older law (Law 48 of 1982) prohibiting the dumping of anything in the Nile or in the canals (see Table 30 below). These figures hide some variation.

Respondents in the three urban areas were about equally likely to say they had heard of the 1994 law (49.8% in Sayyida Zeinab, 48.9% in Kafr al-Elow, 46.3% in Dar al-Salam), about 12–15 percentage points ahead of the people of Abkhas (34.9%). Although a majority in all areas said they knew of the antidumping law, paradoxically the two areas without canals were more likely to recollect the law prohibiting dumping in canals (77.6% in Dar al-Salam, 75.2% in Sayyida Zeinab), while the two localities with canals were more forgetful (68.1% in Kafr al-Elow, 61.5% in Abkhas). Males were more likely to know of both laws (77.5% to 64.4% for the antidumping law, and 51.2% to 39.7% for the 1994 environment law). Knowledge of both laws decreased with age, and increased with education. Overall, while 55.4% of the schooled knew of the new environmental law, rising to 73.1% for those with postsecondary education, only 32.9% of the unschooled said they did. Figures for knowledge of the antidumping law were comparable: 82.5% of the schooled, rising to 93.8% of those with post-secondary education, against 56.6% of the unschooled. It is not surprising that knowledge of these laws is higher among men, the younger, and the more educated, not least because these factors correlate among themselves.

The Next Generation

Intergenerational responsibility was accepted by 95.1% of the 1995 sample; this rose to 98.6% in 1997. There is no strong variation by age, gender, or locality, though the oldest group feels this least (91.4%).

We then asked the people who accepted responsibility to say how that responsibility should be manifested. Table 29 is the breakdown by answer to the open-ended questions.

One can note that several of these are not direct answers to the question, but instead seem to reflect the primary concern of the person answering. Those who answered the question more or less directly stressed the need to teach children, especially about cleanliness and the Quran. The role of education is certainly one possible course of action, but it reflects an essentially passive attitude toward actual intervention. The emphasis in these answers (in spite of the fact that the two largest answers are "passive") is on getting other people to change their behavior, on changing one's own behavior, and on pressuring authorities to act. We may not have elucidated the point of the question sufficiently, which was what the current generation should do to leave a better environment for the next one (the notion of stewardship). Instead people answered in terms of the responsibility of the adult generation for the youth. This reflects their understanding, and their failure to understand our question is itself revealing.

Of those who said there was no responsibility, most (77.1% of 109 cases; one did not answer) said that people are helpless; if officials are not interested, then what can people do? In other words, this harks back to the reliance on government initiative to solve all problems. Another nine (8.3%) said there was no such thing as a clean environment, and the remainder said they had no ideas.

Table 29: Responsibility to the next generation

Responses to the question: "How should we show our responsibility to the next generation?"

	n	%
Know, teach the Quran, create awareness	347	16.1
Teach cleanliness	334	15.5
Clean our house and street	217	10.1
If clean place found, move	161	7.5
Be clean in dress and food	144	6.7
Let each person solve his / her problem(s)	121	5.6
Keep our children away from polluted areas	85	3.9
Get specialized govt. organizations to do some work	77	3.6
Cooperate with other people	67	3.1
Fill in the canals	50	2.3
Build factories far from people, build roads, teach cleanliness	49	2.3
Implement awareness program and laws for factories	48	2.2
Use education + street cleaning + sewage	46	2.1
"Need money to solve problems"*	43	2.0
Live in a healthy place	42	1.9
Clean up street, get people to throw garbage in cans	39	1.8
"The child is a mirror of his environment"*	39	1.8
Install filters on cement factories	30	1.4
Develop planning strategies	26	1.2
Solve the sewage problem	24	1.1
"It should be everyone's responsibility"*	17	0.8
Clean the streets and plant trees	9	0.4
Increase green areas	8	0.4
Use new technology to prevent car exhaust	4	0.2
Feel responsible but no definite recommendations	129	6.0
Total of those who accept some responsibility	2,156	100.0
Do not accept responsibility	109	
No answer	1	
Grand total	2,266	

* Indirect answer, but reflecting the primary concern of respondent.

Source: 1995 survey. Percentages are of those who accepted there was some intergenerational responsibility for the environment (n=2,156).

Media, Knowledge, and Attitudes

Overall, 80.1% of the 1995 sample identified mass media (all types) as one of their main sources of information about the environment,[72] followed by learning from relatives and friends (15.4%), from personal observation (11.3%), and from school courses (9.3%). These figures were nearly the same for women and for men. They varied by locality. People in Kafr al-Elow were least likely to cite the media as a source (65.5%), and most likely to mention both personal observation (19.2%) and relatives and friends (19.1%). People in Sayyida Zeinab were most likely to cite the media (91.8%) and school courses (12.9%).

These figures for learning from the media may seem exaggerated, but they are consistent with other survey information. It may be that it is culturally acceptable in Egypt to cite the media as a source of information. On the other hand, the consistent differences between the four research localities suggest the role played by specific local circumstances. Hannigan (1995:24) phrases the point the other way around: "Public concern is at least partially independent of actual environmental deterioration and is shaped by other considerations; for example, the extent of mass media coverage."

Among the media, reading a newspaper seems to inform people better than watching television, despite what people say about themselves (see Tables 30 and 31), and it seems to affect people's state of knowledge more than their opinions. Thus, for each of the six questions in Table 30, television watchers are only just more familiar with envirnomental topics than the whole sample, while newspaper readers are noticeably better informed. (This could be a result of the fact that 91.3% of the sample say they watch television, while 45% say they read newspapers.) However, the gap between newspaper readers and the aggregate figure is larger when it is a matter of knowledge (questions 1–4) than when it is a matter of opinion (questions 5–6).

It is also worth pointing out that while those who watch three hours or more of television a day gave more positive answers to all six questions than the overall sample, those who watch two hours or less gave more negative answers than the overall sample to the last three questions on the list, including the two attitude questions. In other words, low watchers score lower in familiarity with environmental topics than either high watchers or newspaper readers, though the differences are small.

This is indicated in Table 30, where "readers" are newspaper readers and "watchers" are those who say they watch television. Under question 5 we have included both those who said "something" and those who said "eveything.". Under question 6 we have included both those who said people were somewhat concerned and those who said they were fully concerned. Under question 4 we have included all those who gave some kind of an answer, i.e., we have excluded only those who said they did not know.

Table 30: Effect of mass media on knowledge and attitudes

*Positive answers to question**	*All*	*Readers (%)*	*Watchers*
1	45.5	61.3	47.0
2	71.1	86.1	73.5
3	22.4	30.9	23.7
4	28.8	34.8	29.2
5	40.2	43.3	40.9
6	71.2	72.9	71.7

* The questions were:
1. Do you know the new environment law?
2. Do you know the anti-dumping law?
3. Does the EEAA help protect the environment?
4. Do you know somewhere to go for help?
5. Do you think the government has done something about the environment? (see Table 38)
6. Do you think people are concerned about pollution? (see Table 24)

Source: 1995 survey. Percentages are of the whole sample (N_{1995}=2,266).

Overall 70.1% of our sample said they had watched television spots on the environment.[73] Abkhas was the lowest with 58.3%, then Kafr al-Elow with 66.2% and Sayyida Zeinab with 74.2%, finally Dar al-Salam the highest with 79.3%. Women are more likely to say they watch the spots than men are. The more educated watched them more, or at least noticed them more. Those who said they watched the spots generally found them good because of the information they contained and the fact that they made people aware. Only 5.8% of those who saw the spots qualified them as not effective at all.

Our respondents were more likely to say that television ads changed their habits or attitudes concerning the environment than to attribute a change of attitude to reading newspapers (Table 31). Again the main difference is by locality.

Among the information respondents cited from television ads are generalized awareness (30.4% of the subset of 760; i.e. 32.9% of the total sample), the need to save water (14.6%), that one should avoid throwing leftovers or garbage in the sewage or sink (14.3%), that garbage should be thrown in waste bins or barrels (10.1%), that it is bad to sell uncovered bread in the

Table 31: Media and changing attitudes

Respondents who thought that habits/attitudes concerning environment and pollution had been changed by television ads or newspaper articles

		Locality			
	All	*KE*	*DS*	*SZ*	*AB*
Medium		*(%)*			
By TV ads	32.9	47.8	36.1	32.7	11.9
By newspaper articles	5.5	7.2	7.0	4.2	3.2

KE=Kafr al-Elow, DS=Dar al-Salam, SZ=Sayyida Zeinab, AB=Abkhas.
Source: 1997 survey. Percentages are of the whole sample (N_{1997}=2,307).

streets (9.1%), that is important to eat the right kinds of food and remain clean (7.8%). A total of seven individuals, drawn from each of the four localities, concluded from the television ads that people should not smoke.

Those who reacted to newspaper articles mostly stressed the particular information (articles about chemicals or lead, etc.) they learned there (64.1% of the subset of 128), or said generally that the articles added to their knowledge about cleanliness (14.1%). A group of nine (7%), of whom six were from Kafr al-Elow, mentioned that they learned from newspapers about various conferences on environmental problems.

Our evidence shows that people attribute a good deal of influence to television, but that newspapers also have an impact, and sometimes seem to contribute more solid information. Respondents say that direct observation also plays a role, and the differences between the localities tend to confirm that role. The behavior affected by the media and by direct observation mostly has to do with maintaining cleanliness in one's surroundings.

Conclusion

The respondents generally have a clear idea of what pollution is, at least insofar as their own lives are affected, and they are concerned about it. This is one type of Egyptian environmentalism. Respondents react more precisely to the notion of pollution than to that of environment, which is a more ambiguous term. The sense of nature does not much enter into these definitions, in contrast to the U.S. situation. Their concerns about pollution can be ranked in order of concern from garbage and sewage (dirty streets) to air pollution, water pollution, and noise pollution, with other possible types ranking lower. Our respondents feel that the poor are more affected by pollution than the rich, the very young and very old more than those in the prime of age. Despite a strong sense of the problems caused by pollution, the respondents generally describe themselves as optimistic, and seem to have become more so between our two surveys in 1995 and 1997. They believe that religious teaching is relevant to the environment mostly because it prescribes personal cleanliness before prayer. They describe themselves as influenced by the media, though some of the evidence also suggests the importance of the actual situation in which people live. Newspapers have as much influence as television.

Our respondents and participants feel that the behavior of other people, more or less like themselves, is the major cause of pollution, ahead of industry. In this, the stereotypical case is doubtless the neighbor who discards his waste in a public place such as the street. This tendency to blame other people means in turn that most of the remedial actions the respondents can envisage have to do with correcting the behavior of these irresponsible others, as we will examine in chapter seven.

6

Health

> First comes the problem of garbage. This is the basic problem. Everywhere you go you find garbage. This, you know, is very dangerous to health. Second comes the air pollution problem. The air we inhale every day is dangerous.
>
> Female student, Cairo University, 1993

> I live in fear that smog will choke my children. My youngest son is suffering from a chronic cough and bronchial asthma. He spends most of his time in bed, ill.
>
> Mother in al-Waily, quoted in Kamel 1994:31

Health is a critical issue in the understanding of the quality of life. In this chapter we examine perceptions and understandings relating to health in the context of the quality of life in our research areas. In many respects, health is related to the patterns of everyday living (Tekçe et al. 1994:15), and to the systems of waste disposal described in previous chapters. "Where water supply and waste disposal are deficient, the sanitary discipline necessary to minimize the risks of exposure places enormous daily burdens on household members" (p. 142). Although diagnostic interpretation of symptoms given in a questionnaire is uncertain, we present some data based on self-reporting.

Our data represent an element in the cultural construction of health rather than an epidemiological analysis. We include the results from the 1995 survey with those from the 1997 survey and the focus groups.

Evaluations

We asked people how they rated their health, and gave them a choice of five answers: excellent, good, fair, bad, and very bad. The discussion here is based on the percentage in each category who described their health as either good (the larger number) or excellent (Table 32). As an accurate description of health this is certainly wanting, but the variation in the answers tells us something about how different categories perceived their situation. The percentage of those overall (N=entire 1995 sample) who thought their health was good or excellent was 69.5%. However, there was a range between the different localities. People had the most positive sense of their health in Abkhas, where 78.3% of the respondents described their health as good or excellent. In fact, 19.6% described their health as excellent. People in the two inner city areas of Cairo also rated their health highly, with 73.1% in Sayyida Zeinab and 72.2% in Dar al-Salam so describing their health, while in Kafr al-Elow, heavily affected by pollution, only 56.3% of the respondents described their health in such favorable terms. The younger groups have a more positive sense of their health, and the figure drops off rapidly after age 45 (of those 25 and under, 85.4% esti-

Table 32: Health ratings

How do you rate your health?

		Locality			
	All	*KE*	*DS*	*SZ*	*AB*
Rating		(%)			
Excellent	10.9	8.0	6.4	11.5	19.6
Good	58.6	48.3	65.8	61.6	58.7
Fair	26.7	38.1	24.1	23.6	19.6
Bad	3.7	5.5	3.5	3.3	1.9
Very bad	0.1	0.2	0.2	—	0.2
Excellent & good	69.5	56.3	72.2	73.1	78.3

KE=Kafr al-Elow, DS=Dar al-Salam, SZ=Sayyida Zeinab, AB=Abkhas.
Percentages have been rounded to the nearest tenth of a percent.
Source: 1995 survey. Percentages are of the whole sample (N_{1995}=2,266).

mated their health as excellent or good; of those over 45 only 50.7% did so). There is a significant gender difference, with men estimating their health as excellent or good in 73.2% of cases and women in 65.5% of cases.[74]

Overall, the categories of locality and level of education demonstrate the greatest range of variation in the answers to these questions. In terms of localities, the four sites approximate three positions. The people of Kafr al-Elow have the lowest estimation of their health, while those of Abkhas have the highest, and the two urban sites are in between. As in the overall health picture, in these three subsets men are less likely than women to feel a problem. The older age groups have a more negative picture of their health. And by and large the higher the level of education, the fewer health problems people perceive. Thus, among the illiterate, 65.7% said they had no serious disease by their own definition, while among those with postsecondary education, the figure was 87.4%. Those with some schooling are substantially less likely to complain of headaches, of having trouble going to sleep, or a serious disease than those without schooling.

Assuming that lack of education is correlated with poverty, this self-evaluation is consistent with the observation that the health of the poor is more problematic than that of the better-off. It should, however, be kept in mind that many of the unschooled are women who generally have a lower estimate of their health.

Table 33: Health experience of respondents

Symptom experienced	*Never (%)*
Headache	49.7
Sleeplessness	63.2
Irritability, depression	65.0
Difficulty waking up	75.8
General weakness	76.0
Breathing difficulty	78.2
Irregular heart beat	82.3
Skin irritation	85.3
Dizziness, fainting	85.4
Swelling	93.5
	No
Do you have a serious disease?	76.6

Source: 1995 survey. Percentages are of the whole sample (N_{1995}=2,266).

We asked respondents if they had ever experienced any of a range of symptoms (possible answers: always; sometimes; never). Table 33 shows the percentages of people who answered "never" to the symptoms cited. The ranking in the table is thus from the most common complaint to the least common.

To develop this analysis, we have focused on the two most commonly cited ailments in the overall sample: headache pain, and trouble going to sleep. As in Table 33, we focus on those who said they never had the problem, and we add an analysis of those who said they had no serious disease.

Table 34: Respondents' health experience by locality, gender, and education

		Locality				*Gender*		*Formal education*	
	All	*KE*	*DS*	*SZ*	*AB*	*F*	*M*	*None*	*Some*
				(%)					
No headaches	49.7	38.3	47.3	43.8	75.1	46.9	52.5	40.8	59.2
No sleeplessness	63.2	47.6	57.6	63.6	89.6	61.2	65.1	40.3	59.7
No serious disease	76.6	69.4	78.6	76.5	84.0	72.1	81.2	38.3	61.7

KE=Kafr al-Elow, DS=Dar al-Salam, SZ=Sayyida Zeinab, AB=Abkhas.
Source: 1995 survey. Percentages are of the whole sample (N_{1995}=2,266).

Among those who complained of a serious disease, the most commonly cited problems were: chest trouble (4.6% of the total sample), bone trouble (4.4%), diabetes (3.3%), blood pressure (2.6%), skin diseases (2.3%), heart trouble (2.2%), eye trouble (1.9%), kidney trouble (1.5%), stomach trouble (1.2%). All others were mentioned by less than 1% and included liver trouble, bilharzia, hookworm *(Ancylostoma duodenale)*, typhoid, malaria, anemia, headache, ear trouble, nerve trouble, varicose veins, foot trouble, tooth trouble, sinus problems, brain trouble, and old age. Altogether 528 people (23.3%) cited one or more diseases for a total of 629 answers.

Reported Illnesses in 1997

In 1997, people generally were even more positive about their health. Overall 12.7% described their health as very good or excellent, and 61.3% as good, making a total of 74% compared to 69.5% two years earlier. Only 26% described their health as fair, bad, or very bad. As in 1995, the lowest overall estimate of health was in Kafr al-Elow and the highest in the village of Abkhas. However, despite this overall high evaluation, 22.1% of the total sample reported that they had been sick in the last two weeks, and about three quarters of those felt they were sick enough to have missed work. The most common complaints were: rheumatic, muscle, and bone pains; liver, kidney, or gall bladder problems; chest ailments or lung infections; high blood pressure; diabetes.

All these ailments were reported more commonly in the three urban sites than in the village. We note that males are much more likely to report (or have reported for them) the symptoms in question although females seem to complain of worse health overall; there appears to be a gender bias.

Of families with children (1,306 of a total of 2,307 households, or 57%), 13.6% reported diarrhea among these children within the previous two weeks, and 9% reported respiratory illnesses.[75] These illnesses affected primarily the younger children. Nearly half (47.7%) of the reported cases of diarrhea involved children aged 2 years or less. Just over half (52%) of those reporting respiratory ailments were aged 5 years or less. Taking all family members into account, 13.9% reported chest allergies, 7.7% reported skin allergies, and only 0.7% reported asthma. Even though the age range for those affected with chest allergies was from 1 year to 85 years, 21.4% were aged 5 or less, and the median age was 17. In the case of asthma, the age range of this small group was smaller, from 1 year to 73 years, with a median of 60 years. All these symptoms were reported more often by or on behalf of males than by or on behalf of females.

About a quarter of the 1997 sample (27.4%; n=633) reported having had an operation. The most common were operations on the stomach (27% of the subset), gynecological operations (19.7%), nose, ear, or throat operations (16.7%), and urinary or bladder operations (15%).

When people (including their children) suffer from a disease, they overwhelmingly consult Western medicine in one form or another—either private doctors or government hospitals. Others consult private clinics or religious charity clinics. Only a few try traditional remedies (in the case of infant diarrhea, 8.5% of those reporting a case used [unspecified] traditional remedies, while another 6% did nothing).

Pollution and Health

In a series of questions, we asked the 1995 respondents whether they felt that polluted air, polluted water, or noise pollution had a negative effect on their health, on the health of their family members, and on their quality of life in general. People were asked to give one of four answers: strong effect, moderate effect, weak effect, and no effect. As the answers to these various questions were largely consistent, we have summarized them by citing only the percentage of those in each category who felt that air, water, or noise pollution strongly affected their own health. These figures are given in Table 35; it can be seen that the figures are generally fairly high. As with many of our results, the relative value may be more significant than the absolute value.

From Table 35 it appears that people are more worried about air than about water, and about water more than about noise. This pattern is true for each of our four localities, though the spread between the figures varies. It is noteworthy however, and also generally consistent with other results, that the

Table 35: Perceptions of the effect of pollution on health

Which of the following (forms of pollution) has a strong effect on your health?

Form		*Locality*				*Gender*	
	All	*KE*	*DS*	*SZ* (%)	*AB*	*F*	*M*
Air	78.1	**85.9**	80.6	81.9	60.0	77.3	78.8
Water	63.6	49.3	**80.1**	75.5	45.7	61.3	65.7
Noise	46.3	29.1	59.1	**67.6**	25.1	46.1	46.6

KE=Kafr al-Elow, DS=Dar al-Salam, SZ=Sayyida Zeinab, AB=Abkhas.

Highest score for each row in bold.

Source: 1995 survey. Percentages are of the whole sample (N_{1995}=2,266).

locality whose inhabitants worry most about air pollution affecting health is Kafr al-Elow, while Dar al-Salam inhabitants worry most about water pollution and Sayyida Zeinab inhabitants worry most about noise. Abkhas has the lowest percentage of people who feel that each of the three has a strong effect on their health.

Level of education is clearly reflected in the answers. Those who answered that air pollution strongly affected their health were 75.4% of the illiterates and 84.9% of the university graduates.[76] For water pollution, the spread was greater: 57.2% of the illiterate and 78.4% of university graduates, and for noise pollution, it was greater yet: 35.7% to 69.4% respectively. Regardless of the type of pollution, the number of respondents thinking it affected their health rose progressively with increasing level of education. The level of concern also rose slightly with age, reflecting a general increase in concern with health as people age. The youngest group shows the lowest level of concern for each of the three forms of pollution, but the group most concerned about water and noise is the second oldest quartile (those aged 36–45 years), though the two older groups are within about a percentage point of each other. Gender differences are minimal as well, the biggest difference being that 65.7% of the men and 61.3% of the women feel that polluted water is a serious health threat (see Table 35).

The Kafr al-Elow focus group participants often mentioned the health consequences of air pollution—asthma and "chest allergies," eye infections, kidney problems. There is a general feeling that the generation now growing up is not as healthy as the older generation who grew up in cleaner times. According to one participant, "Now somebody who is 60 years old, but who grew up and was nourished on food during the good old days, has more energy and is healthier than a 20-year-old person." Participants say children are malnourished and weak, and they look older than their age. The elderly also suffer from air pollution. However, today's adults are not immune, and some women report general fatigue as a result of the pollution. Women in particular mention that their doctors advise them to move from Kafr al-Elow in order to recover their health.[77]

A public-sector battery factory is located in Dar al-Salam. It had been closed for about two years when we interviewed a group of its workers (who were still receiving their pay) in January 1996. The workers knew that their job was dangerous, and were just as happy to be paid without having to work. Workers in the factory had gone through periodic blood tests to ensure that the level of lead in their blood was not a mortality risk. However, all of them have higher lead levels than normal. They recognize the symptoms of lead poisoning as general fatigue, a constant desire to sleep, lack of energy, loosening teeth and tooth loss, a yellowish face, and bluish lips. In addition to lead they were also exposed to other heavy metals.[78] Because of the lack of any industrial safety measures at the factory, and because the factory was located within a residential area, the effects of the lead were very serious both for the workers and for the residents (those living in a radius of 200 meters). The factory managers, according to the workers, had no concern for health measures, but were corrupt and careless. Individual lawsuits were filed against the factory without result, though the factory closed.

The idea is fairly widespread in Egypt that people suffer kidney failure and other diseases because of the polluted drinking water. One of Egypt's leading ecologists and environmentalists, Dr. Abdel-Fettah Al-Qassas, is quoted in *Al-Ahram Weekly* of June 5, 1997 (Nasr and Bakr 1997a), as saying that kid-

Table 36: Pollution and health, by major background factors

Percentage of respondents who said that polluted air strongly affects their health (strong effect), contrasted with those who rated their general health excellent or good (good health).

	Strong effect (%)	*Good health*
Overall	78.1	69.6
Gender		
Male	**78.8**	**73.3**
Female	77.4	63.7
Educational level		
—No formal schooling		
Illiterate	75.6	57.8
Can read and write	78.6	63.5
—Schooling completed		
Primary	78.2	66.1
Preparatory	72.5	76.1
Secondary	78.2	**82.0**
Post-secondary	**86.7**	80.8
Age (years)		
≤25	72.0	**85.4**
26–35	77.3	75.0
36–45	**81.2**	70.0
≥46	77.7	50.6
Occupation		
Professional, technical, etc.	**90.8**	81.5
Administrative, executive, managerial	77.8	79.9
Clerical workers	67.1	86.8
Owners of businesses	77.6	70.1
Farmers, agricultural workers, fishermen	61.9	73.0
Industrial laborers, craftsmen	81.1	73.0
Unskilled labor in factories	83.0	72.6
Retired	80.3	40.1
Housewife	76.7	62.2
Student	79.0	**88.8**

Bold = highest in each group.

Source: 1995 survey. Percentages are of the whole sample (N_{1995}=2,266).

ney failure has become common in young people because "large quantities of hazardous industrial, agricultural, and domestic wastes are thrown into the Nile and the water has become seriously polluted."

Table 36 analyzes the somewhat contradictory results of two questions, with the answers broken down by gender, locality, education, age, and occupation. The proportion of those who feel that polluted air strongly affects their health is greater than that of those who feel that their health is excellent or good (78.1% to 69.6% respectively).[79] In other words, nearly half, at least, of the sample agrees with both statements.

Men are both more likely to feel that air pollution affects their health and more optimistic about their health. The more educated people are, the more likely they are to worry about air pollution but feel confident about their health. The middle-aged worry more about air pollution, but concerns about health decline with age. The occupations requiring more education, including student, follow the education picture.

Sports and Smoking

Being active in sports and not smoking can be taken as signs, albeit indirect, of concern for health (see Table 37). The figures for smoking and practicing sports show a strong gender bias.

Overall, about a quarter of the sample (24.3%) admits to smoking, including about half the men and a handful of women, which is an enormous gender disparity (even allowing for the possibility that women were more reluctant to say that they smoked). Another noteworthy point is the relatively high percentage of smokers overall. Abkhas has the highest figure, pulled up in part because of the higher relative number of males in the sample there.

Those who play sports are 9.5% of the overall sample. The main sports are soccer for men, and walking and exercising equally for men and for women. Men also mention weightlifting, body building, swimming, and karate and kung fu, while wrestling, boxing, volleyball, basketball, and gymnastics are equally practiced by men and by women. The figure for those who play sports can be compared to the 1995 figure of 6.3% who belonged to a sporting club.

Table 37: Sports and smoking habits of respondents, by locality and gender

		Locality				*Gender*	
	All	*KE*	*DS*	*SZ*	*AB*	*F*	*M*
Respondents who:				*(%)*			
Play sports	9.5	11.6	8.0	8.2	10.3	4.5	14.3
Smoke	24.3	20.1	23.3	19.9	35.8	0.3	47.4

KE=Kafr al-Elow, DS=Dar al-Salam, SZ=Sayyida Zeinab, AB=Abkhas.

F= female, M=male.

Source: 1997 survey. Percentages are of the whole sample (N_{1997}=2,307).

Health Care and Insurance

About a third (33.7%) of our 1997 sample is covered by medical insurance, mostly (67.6%) as individuals only. Men were significantly more likely than women to have health insurance (47.9% to 19.3% of respondents respectively). The locality with the highest proportion of insured respondents is Abkhas (42.5% of the inhabitants), while the urban areas range from 30.4% to 32.6%. This pattern seems to reflect the link between health insurance and employment.

Most people seek medical help from a private clinic (54.2%), while 20.5% go to a public hospital, 10.3% an insurance clinic, 8.9% a religious charity clinic, and 3.6% a private hospital. Respondents in Sayyida Zeinab and Dar al-Salam are more likely than average to use a private clinic, while those in Kafr al-Elow and Abkhas are more likely to go to a public hospital. The religious charity clinics are almost unknown in the village, but are used about equally by the three urban sites. Women are slightly more likely than men to seek help both from the private clinics and the public hospitals, while men, because they are more likely to be employed, are much more likely to seek help from the insurance clinics.

Conclusion

When it comes to health, people are worried about air, water, and noise in that order. Somewhat paradoxically, people feel good about their health in general but at the same time have a lot of specific complaints. They link some of these complaints to environmental circumstances, though it is hard to confirm their feelings scientifically. Among diseases with probable environmental etiologies, diarrhea is reported more often than respiratory problems. Despite general levels of concern about health, people are more likely to smoke than to exercise.

The data in this chapter show that people are aware of the possible ill effects on health of the pollution they experience, even those who feel confident that their own health is good. In terms of the evaluation of their own health, the data show men more than women, and the schooled more than the unschooled (of course these two variables overlap), are confident that they are healthy. People report a variety of symptoms that could be related to exposure to pollution; however, there is no proof offered here of this either on an individual or an epidemiological basis.

7

The Politics of Environmental Action

> Nobody is doing anything to solve the problem. The government is not doing anything. Maybe the government is doing something in other places, but where I live, nothing is being done. We went to the local authorities more than once and filed complaints. I went with some of the men who live in the area and asked [the authorities] if they are going to install sewage pipes or not, and when. They said that they will install sewage in the proper time, that the district is part of the plan, etc. But we do not believe them.[80]
>
> Dar al-Salam taxi driver, 1993.

This chapter is devoted to identifying the forms of action that people in the sample areas considered or undertook to improve their environmental conditions. Since many people felt that the most significant actions could be taken only by the public authority, we start with feelings about government responsibility, asking what people perceive the government to have done, and what it can be expected to do. We then turn to the feelings that people have about cooperating with their neighbors, and dealing with neighbors whose behavior is bothersome to them. Finally, we present material relating to three forms of political action: complaining to the authorities, contacting authorities to persuade them to carry out a course of action, and lastly voting as a way of putting pressure on elected officials to endorse certain policies. People are also aware of the possibility of direct action or of lawsuits.

There is no simple correlation between thought and action. Nevertheless, we have argued that thought (the social and cultural constructions of a given situation) provides a frame for action. How people define a problem determines what solution they seek, and hence the tendency of the Egyptians in our sample to blame others for pollution means, at face value, that they seek to improve the situation by correcting the individual behavior of others like themselves.

The Role of the Government

> I believe the government is responsible for the situation in Dar al-Salam. It is a complete failure. Every now and then they say things they can not fulfill or do. I believe that our government is responsible because it chooses to pave Heliopolis streets three times a year and decides not to pave any street in Dar al-Salam. This is the bias that we suffer from.
>
> Dar al-Salam taxi driver, 1993

Most of our respondents thought the government has done nothing to protect and clean up the environment (Table 38), but there was some variation

Table 38: Perceptions of the government's environmental role

Has the government done enough to protect and clean up the environment?

		Locality			
Government has done:	*All*	*KE*	*DS* (%)	*SZ*	*AB*
Everything	13.6	12.6	17.5	16.0	7.0
Something	26.6	20.8	33.5	30.8	19.8
Nothing	57.5	64.2	45.5	50.7	72.8
Other (specify)*	2.3	2.3	3.5	2.5	0.4

* See Table 40

KE=Kafr al-Elow, DS=Dar al-Salam, SZ=Sayyida Zeinab, AB=Abkhas.

Source: 1995 survey. Percentages are of the whole sample (N_{1995}=2,266).

Table 39: Perceptions of others' concern for the environment

Do you think that those around you are concerned about the environment?

		Locality			
	All	*KE*	*DS*	*SZ*	*AB*
Level of concern			(%)		
Very	28.0	32.6	37.4	20.8	19.6
A little	43.2	40.8	39.1	49.3	43.6
None	28.4	26.3	22.6	29.8	36.8
Don't know	0.4	0.3	1.0	0.2	—

See also Tables 24 and 40

KE=Kafr al-Elow, DS=Dar al-Salam, SZ=Sayyida Zeinab, AB=Abkhas.

Columns slightly exceed 100% due to rounding.

Source: 1995 survey. Percentages are of the whole sample (N_{1995}=2,266).

from one locality to another. Only in Dar al-Salam did a majority of people feel that the government had had a positive effect on the environment (51% of respondents, against the 45.5% who thought the government had done nothing). In each locality, those who thought the government had done the minimum (but could do more) outnumbered those who thought it was doing all right.

Tables 38 and 39 show that those who thought others were very concerned about the environment numbered about twice those who thought the government had done everything it could, though in turn people were more likely to think that they themselves were concerned about the environment than to think that others were (see Table 40).

Most people felt that those around them were at least a little concerned about the environment, though not necessarily very concerned (Table 39). Dar al-Salam and Kafr al-Elow lead the category of those who think their neighbors are very concerned, and Abkhas and Sayyida Zeinab lead those who think their neighbors are not concerned; this corresponds approximately at least to the actual circumstances our study revealed in those localities (worse in Kafr al-Elow and Dar al-Salam, better in Sayyida Zeinab and Abkhas).

Table 40 offers a comparison between the responses we received to four questions in 1995, which we have already examined separately above. Two questions have to do with people's evaluation of the role of others in protecting or cleaning up the environment: that of the government and that of

Table 40: Opinions about the government, others, oneself

	*Respondents agreeing with statement**			
	1	*2*	*3*	*4*
		(%)		
Overall	57.5	28.4	50.4	79.8
Locality				
KE	64.2	26.3	59.4	72.2
DS	45.5	22.6	53.5	85.4
SZ	50.7	29.8	50.4	81.9
AB	72.8	36.8	34.7	80.0
Gender				
M	55.0	27.9	53.0	79.6
F	60.7	29.0	47.6	80.1
Age (years)				
≤ 25	59.5	32.4	54.4	80.6
26–35	59.9	26.7	49.9	78.9
36–45	53.8	26.0	52.0	81.9
≥ 46	57.7	29.5	46.7	78.6
Educational level				
—No formal schooling				
Illiterate	59.3	29.8	35.1	76.4
Can read and write	62.3	29.0	47.0	79.2
—Formal schooling completed				
Primary	59.2	21.3	51.1	82.2
Preparatory	59.9	20.8	52.1	85.9
Secondary	46.6	29.3	53.4	88.0
Technical	60.6	28.8	61.8	78.6
Post-secondary	49.3	32.2	64.4	80.4

*Statements were: (1) The Government has done nothing to protect or clean up the environment (see Table 38); (2) Those around me are not concerned about the environment (see Tables 24 and 39); (3) I am very concerned about the environment (see Table 8); and (4) I am an optimist with regard to the environment (see Table 28).

KE=Kafr al-Elow, DS=Dar al-Salam, SZ=Sayyida Zeinab, AB=Abkhas.

F=female, M=male.

See also Tables 8, 24, 28, 38, 39

Source: 1995 survey. Percentages are of the whole sample (N_{1995}=2,266).

those around them. Two other questions are self-evaluations of the respondents' level of optimism and level of concern regarding the environment. For each of these questions, the answers are broken down by locality, gender, age, and level of education.

The majority of people felt that the government had done nothing to protect and clean up the environment.[81] This majority extended to every category except, marginally, those with a secondary and a post-secondary education. This feeling was particularly strong in Abkhas, which is otherwise the community least concerned about environmental issues (for instance, it has the most respondents who call themselves optimists), though it was predictably strong in Kafr al-Elow as well. The strength of this figure in Abkhas points to a problem in the data: almost certainly some, perhaps many, of the people who gave this answer feel that the government does not do enough about anything, and this colored their answer so that it reflects the general feeling rather than an evaluation of government efforts in the environmental area (see the focus group data). Although the result reflects considerable alienation, it does not discriminate between these two explanations.

So we tried again in 1997. Only one fifth of the 1997 sample (20.3%) felt that the government has done its duty in cleaning and protecting the environment. This group was most numerous in Dar al-Salam (26.6%) and Sayyida Zeinab (20.5%) followed by Kafr al-Elow (18.6%) and Abkhas (14.6%). Though the actual figures were different, the ranking of the four localities was identical to that two years earlier. The most commonly cited positive activities of the government had to do with waste disposal—34.5% of the people who felt the government had done its duty cited managing garbage, while another 26.3% cited installation of a sewage system (in fact a sewage system was being built or modernized in Dar al-Salam and Sayyida Zeinab during the period of research). Another 11% mentioned the government's role in conducting awareness campaigns. Smaller numbers cited establishing parks, installing filters in factories (Kafr al-Elow), paving roads (but none in Abkhas), moving workshops from residential areas (in Abkhas). Three individuals cited the creation of the EEAA.[82]

The gender data in Table 40 show that men were more positively inclined toward government efforts than women. In terms of age, the group aged from 36 to 45 years was the most positively inclined. As we have noted, those with secondary and post-secondary education were the most inclined to feel the government was doing something, though not necessarily enough. In sum, those most favorably disposed toward the government tended to be middle-aged men with a good education.

Overall just over a quarter of the 1995 respondents (28.4%; see Table 9) felt that those around them were not concerned about the environment. This measure of alienation was most common in Dar al-Salam and least common in Abkhas; it was about equal among men and women; it was more common among young respondents, and among those at the two ends of the educational spectrum. But whereas the figure for self-identification as environmentally concerned rises consistently with educational level, the figure for estimating the concern of others does not. People are consistently more likely to see themselves rather than others as concerned about the environment. This can be taken as evidence of a sense of isolation when it comes to environmental issues.

By and large, those who feel that people around them are concerned about the environment are somewhat more likely to think that the government has done a good job in dealing with the environment. Thus those who think that people around them are very concerned with the environment are 28% of the total sample, but they are 51.1% of those who think that the government has done a good job. Conversely, those who think that people around them are unconcerned are 28.4% of the total sample, but they are 34.2% of those who think that the government has done nothing. In other words, there is a tendency for these two judgments to run in the same direction. This could be taken as an indicator of the level of interest of the respondent: those interested in the environment think that others are, and also think the government is active.

These points were elaborated by participants in the focus groups. These participants did not see the government (often meaning local government) as active or effective. They expressed a yearning for a government that would establish and strictly enforce clear laws. Some said this existed in the past, but no more. People would like leaders who set them an example and assert direction, and then they would be prepared to join in. They sought synergy not disengagement. At the same time, they also do not see easily how they can cooperate with each other, given people's preoccupation with personal economic matters, the difficulty of cooperation, increasing individualism, and general passivity and apathy.

People often approach the local council *(maglis al-hayy)* in Kafr al-Elow, but without much effect. The council seems uninterested, but is also hampered by bureaucratic rules and procedures, routine, and inefficiency, and apathy on the part of the people. In any case, Kafr al-Elow is too far down the government hierarchy, and doesn't get much attention. In fact, one suggestion in the focus groups was to lobby to make Helwan into a separate governorate, so that officials would pay attention to local problems that are often overlooked in the much larger Cairo governorate.

Dar al-Salam participants believed that government officials would not do anything and that it was a waste of time to contact them. They complained about their local council, which they said lacks the ability to enforce rules and often does not answer complaints. People approached the council to complain about the condition of the streets, street garbage dumps, lack of lighting, potholes, open drains/sewers, workshop noise, etc. The response of council officials was that they could not do anything to help, being understaffed. Nevertheless, residents felt that because they pay taxes they are entitled to services, as much as are people in fancier neighborhoods. Even if the pressure to support their families means that they do not have a lot of time to spend pressuring the local council to live up to its job, people continue to look to the council for help. For instance, it was suggested that the local council could sign a contract with a reliable garbage collector, and could then enforce the contract, which individuals are in no position to do.

In Sayyida Zeinab, focus groups noted that many problems stem from government neglect. The government is responsible for cleaning the streets and collecting the garbage, and it does not do this adequately. The people's role lies in sorting the garbage and in economizing on water. People are dubious that approaching the government to do its part helps much. They feel that government employees are likely to say that it is not the government's business, and so the government and its representatives do not respond to the demands of the people. Government and people blame each other. Failure to get the government to respond has made people apathetic and passive. However, in the new cities such as Sixth of October, residents and local government take more interest in maintaining a clean environment, and they work together.

Some former residents of Sayyida Zeinab felt that people were responsible for polluting their own environment through their actions, while others felt that they had no alternative. In the latter case, the solution lies in the hands of the government. A former resident of Sayyida Zeinab taking part in the focus group stated, "The problems experienced in the old neighborhoods (sewage, garbage, overpopulation) are perceived as government-caused and government-solved."

Participants in Abkhas were not optimistic either about government efforts or about the success of public collective action. They often spoke of the need for some form of government mobilization, which they found lacking, but did not agree on whether the local government or the national government should take the lead. Government intervention is also essential where the project is costly, as in building a sewage system or paving the roads. But some (from the Young Women's focus group) said that "a substantial part of environmental degradation is caused by lack of government care (i.e., infrastructure and services) for the rural setting," and added that efforts to act are in fact thwarted by the government, although ideally there should be cooperation—the people ask and the government provides infrastructure and enforces laws.

Abkhas participants felt that elected officials mostly talk, and should act instead. In particular, some Abkhas women noted that the members of parliament should stop favoring al-Bagur and other centers, and devote some resources to outlying villages. Thus participants pointed out that al-Bagur town, a few kilometers away, and with a powerful local politician, has both a sewage system and a reliable garbage collection system.

Several participants in Sayyida Zeinab stressed that the government should establish and enforce rules. Officials should fine those who throw their garbage in the streets. There should be supervision and punishment. There should be no discrimination or exceptions in law enforcement, and the law should be enforced equally, not only against the underdogs. Participants both in Sayyida Zeinab and Dar al-Salam cited the case of the Cairo underground Metro as an example of what the government can do when it sets and enforces rules, where the prohibition of smoking is strictly and universally enforced. Similarly, neighborhoods or villages should receive equal treatment in both the application of the law and pollution control initiatives. As in Dar al-Salam, participants complained that they pay taxes according to the rates set for more affluent areas but do not receive the same services.

Many Sayyida Zeinab participants expressed the opinion that collective actions were more common in the 1960s, when government action (and the work of the Socialist Union, then the single political party) was more effective in keeping the streets clean. When people saw the government trying to keep the streets clean they were more careful themselves. "In the old days, the government representatives would go around with a microphone and tell the people that this area was going to be cleaned, and that anyone who broke the [no littering] rule would pay a fine. It worked and people followed the rules," commented one of the members of the Senior Men's focus group. (These themes were most common in this old urban locality.)

In Kafr al-Elow focus group participants suggested that WHO monitor the government, to ensure compliance with national and international standards. The EEAA was not strict enough in enforcing its own rules. In general there was support for the idea of using fines to enforce behavior. But this implies that the government is prepared to enforce certain standards. And somewhat cynically it was pointed out that "since the government owns the factories, it does not necessarily want to solve the pollution problem."

Knowledge and Responsibility

> Some problems are people's responsibility, others are government's, and others are shared. Yet as a rule you may say that people are the major responsible party—if they have not created the problem, at least they should be responsible for pressuring to solve it.
>
> Male student, Cairo University, 1993

One of our concerns was to identify what people thought that individuals like themselves could do, either separately or collectively. In both the 1995 and 1997 surveys we asked a series of questions to probe what people thought they themselves could do, and where they could go or appeal to for help on environmental issues. Their answers partly reflected a level of knowledge or experience, and partly underlying attitudes.

Our 1995 respondents mentioned several things people could do about environmental risk. Almost everyone (83.2%) cited "cleaning the streets," and 48.7% said people should avoid throwing water in the streets. A group of 22.1% thought that avoiding noise-making was important, and 15% said people should avoid smoking where it bothers others, such as in closed-in areas. Various answers could be grouped together under the heading of getting other people to do something (the government, fellow citizens); these amounted to 3.5% of the total.

There is relatively little variation by age or by gender, although again men tended to answer more (multiple answers were allowed). The ranking by locality is the same in each case. Note that the first two answers above are versions of the same idea, to keep the streets clean. Those with higher levels of education were much more likely to cite all of the four main categories, but particularly the avoidance of noise and smoking. They also supported cleaning the streets, but since almost everyone did, the difference is not striking (81% of illiterates to 91.9% of post-secondary levels).

One question was where or to whom people could go if they had an environmental problem (1995; see Table 41). Most said they would go to no one, probably a combination of not knowing where to go and not feeling that anyone in the place where they might go would actually be able to do something. This amounted to 71.2% of the overall sample, with a range of 80.4% in Abkhas, to 70.9% in Sayyida Zeinab, 67.8% in Dar al-Salam, and 67.7% in Kafr al-Elow. In other words, the people in the three urban areas were most likely to think they had somewhere or someone to go to. More respondents in Kafr al-Elow said they would solve the problem themselves rather than go to anyone (10.1%, the highest score), while 7.8% in Sayyida Zeinab, 7.6% in Dar al-Salam, and 6.6% in Abkhas gave this answer, making an overall figure of 8.1%. The others who thought they knew whom to approach mostly cited one or another form of government (16.9%), i.e., the local council under various names, or such institutions as the *omda* or a police station in Abkhas, or a particular company such as the sewage authority (0.5%). Only three respondents mentioned contacting the newspapers to bring publicity, while a few others cited intermediaries such as the *imam* or the *omda*. Another three individuals mentioned the dominant party in Egypt, the National Democratic Party.

Table 41: Seeking help for an environmental problem

Where can one find help for an environmental problem?

		Locality			
	All	*KE*	*DS*	*SZ*	*AB*
			(%)		
No one can help	71.2	67.7	67.8	70.9	80.4
Government office	19.2	20.0	22.7	19.9	12.3
Solve it ourselves	8.1	10.1	7.6	7.8	6.6
Political leaders	0.6	1.2	0.3	0.7	0.4
Other	0.9	1.0	1.6	0.7	0.3

KE=Kafr al-Elow, DS=Dar al-Salam, SZ=Sayyida Zeinab, AB=Abkhas.
Source: 1995 survey. Percentages are of the whole sample (N_{1995}=2,266).

The answers we have grouped under "government office" also include those who mention contacting the office through an intermediary. Those who said they would solve the problem themselves have in mind collecting money from different people in order, for instance, to connect themselves to a sewage system or hire a truck to empty the cesspits. Political leaders include the *imam*, the *omda*, or the local representative in parliament.

Relatively few (in 1995) said they knew of any organization for the protection of the environment (4.5% overall, ranging from 6.5% in Kafr al-Elow to 1.7% in Abkhas). Men were more likely to have heard of one than women (5.5% to 3.4%), young people more than older people, and the most educated more than others (ranging from illiterates at 1.1% to post-secondary level at 13.4%; the jump comes with a secondary level education). Of those who said they knew of such organizations, about two thirds (69.6%) said they would join, although apparently none of them were members. The most frequently identified environmental organization was the Green Party (22 citations from all three Cairo sites).

Table 42 displays the answers to the 1995 question asking respondents to identify which of a list of organizations had the most prominent role in environmental protection. It seems that very little confidence is given either to the political parties or to the NGOs.

Table 42: Perceptions of organizations' role in environmental protection

		Locality			
Organizations thought to have a role	*All*	*KE*	*DS* (%)	*SZ*	*AB*
Local councils	39.4	34.9	41.1	37.1	46.0
Don't know	29.4	36.3	21.5	22.3	39.6
EEAA	22.4	19.3	26.6	32.4	8.1
Political parties	4.6	5.7	5.9	3.2	3.6
NGOs	3.0	2.7	2.9	4.3	2.1
CDA*	1.1	1.2	2.0	0.7	0.6

* CDA= Community Development Association

KE=Kafr al-Elow, DS=Dar al-Salam, SZ=Sayyida Zeinab, AB=Abkhas.

Source: 1995 survey. Percentages are of the whole sample (N_{1995}=2,266).

Apart from those who said they did not know, the answers were divided between those who saw the local councils filling that role, and those who identified the EEAA. The main gender difference here is that men were more likely to cite the local councils as having a role (46.9% to 31.6%), while women were more likely to say that they did not know (37% to 22.1%). On the other options there was negligible gender difference. The older age groups were more likely to cite local councils, while the younger mentioned the EEAA more often, but the differences were slight. In the end, those who think someone can help with their environmental problems rely on the government (local councils, EEAA) to get things done.

Although almost no one mentioned the EEAA in the open-ended question (Table 41), it received more mention when it was listed in a closed-ended question (Table 42). On the other hand, at the end of the questionnaire, when we asked if respondents knew what the role of the EEAA was, only a few could spell it out. The 7.5% who thought they knew generally did know. We conclude from this that in 1995 people had sometimes vaguely heard of the EEAA but they did not have a clear picture of its role.

We asked a similar question in 1997. Of the total 1997 sample, only 97 (4.2%) claimed to know of any political party or NGO that was working to improve the environment. The ratio was highest in Kafr al-Elow (6.5%) and lowest in Abkhas (2.6%). However, of this number, 14 said they did not know what organizations cited did, 15 said they did nothing, and four did not answer. The most common answers among the 64 (66% of the subset) who knew of an organization working to improve the environment and could tell us what it did were that the organization led awareness campaigns (23 cases) or protected the environment from pollution and maintained the greenery (22 cases). Men were more likely to claim knowledge than women (5.6%:2.7% of the total sample); on the other hand, the 31 women who thought they knew something were slightly more knoledgable about what the organization did than the men were (29%:36.4% of the subset).

Table 43 gives the results of a series of five questions (1997) asking what individuals, including the respondent, or various categories or organizations could do to protect the environment. To simplify matters, we have separated the answers into those who felt the entity could do something and those who felt it could do nothing, and since the "don't know" answers are significant, and in some respects may be equivalent to the "nothing" answers, we have listed them separately. The "other" answer includes 20 individuals (10 in Sayyida Zeinab, nine in Kafr al-Elow, one in Abkhas) who noted that the pri-

Table 43: Possibilities for environmental protection

What can these various (nongovernmental) groups do to protect the environment?

	Nothing	*Something*	*Don't Know*	*Other*
		(%)		
Political parties	72.8	1.8	25.1	0.3
Private sector	69.7	7.0	21.6	0.9
"People"	63.7	23.9	12.2	0.2
NGOs	59.5	2.4	32.3	5.6
Oneself	33.4	61.7	2.8	2.1

Source: 1997 survey. Percentages are of the whole sample (N_{1997}=2,307).

vate sector was the worst polluter, and the 129 individuals, in all four localities but mostly in Abkhas, who stated that no NGO at all exists. Arguably, these two answers could also be added to the "nothing" column in that those who articulated them clearly feel that the private sector and NGOs have nothing to contribute.

The main conclusion to be drawn from this set of answers is that people have a hard time imagining what organizations like the private sector, political parties, and the NGOs, could do to help protect the environment. Some of the answers were probably based more on what people knew them to have done than on what they thought they might do; in other words there was some confusion between this question and that detailed in Table 42. On the other hand, people have more idea what "people in general" or "they themselves" might do. Their answers appear to be from experience, not from the discourse of the media.

Some community-based organizations or NGOs existed in Kafr al-Elow, Dar al-Salam, and Sayyida Zeinab, but were focused on charitable activities or on maintaining links between people from the same home town. The environmental NGOs visible at the national level, although they were effective in lobbying for stricter provisions in the 1994 environmental law (Gomaa 1997), do not reach into urban neighborhoods or villages. They are mostly clubs of, at most, several hundred upper class members, and they have neither the staff, the money, the human and financial resources, nor the inclination to take on the problems of the millions of people in poor and middle-class neighborhoods (see M. Bell 1998 and Gomaa 1992).

There is general skepticism among focus group participants on the role of NGOs. In Dar al-Salam, it was pointed out that "NGOs in a poor neighborhood are poor, lacking resources, connections, and a voice. Their activities are limited to charity work, helping orphans, building mosques, managing cemeteries, and so on. These NGOs differ from those found in more affluent areas." Participants felt that approaching them was a waste of time. Lead smelter workers in Dar al-Salam noted that "NGOs, religious institutions, and political parties so far have not played a role in addressing environmental problems. This also applies to the effectiveness of an environment protection law. If this law is not implemented, then it remains useless."

Some Kafr al-Elow women thought that "no NGOs" were active, as they require prestigious people to form them. Community Development Associations (CDAs) and NGOs are elite organizations, to which the people of Kafr al-Elow do not have access. Other people thought that starting their own NGO would be too complicated, so they conceived of the idea of becoming a branch of the Friends of the Environment and Development Association, Alexandria (FEDA-A). They had heard that the FEDA-A was effective in filing and winning cases against industrial establishments. Contacts were being made with this successful association in order to establish a branch in Kafr al-Elow to deal with the cement problem. There is a CDA in Kafr al-Elow, but it does nothing in this area for lack of resources (it receives LE500 a year from the Ministry of Social Affairs), and focuses on a nursery and adult literacy classes.

The idea of forming "block associations" (a form of NGO based on a street or neighborhood) was evoked in Dar al-Salam; but some thought it was not feasible given the underdeveloped status of the neighborhood and the political atmosphere in the country as a whole. Such associations might appear to the government as a kernel of opposition. By the same token, siding with the formal political opposition was seen as too risky.

Taking Action

> The main solution is to make people behave correctly, like in the West. All people have to behave correctly because they know that they will pay dearly if they do not.
>
> Female student, Cairo, 1993

Our respondents generally felt that the main responsibility of individuals was to keep their surroundings clean. Thus 80.4% of those who thought that people could do something about their environmental problems mentioned cleaning the place where they live, getting rid of garbage, and lighting the streets. Others, when asked what people could do to help cited planting trees, starting awareness campaigns (0.5% of the subset), forming organizations in schools, and one person mentioned switching cars to natural gas. Similarly 76.1% of those who could think of something they would do themselves mentioned not throwing garbage in the streets or not burning garbage, or, more positively, cleaning in front of houses, and, slightly more ambitiously, cooperating with neighbors to clean the streets or hiring a back hoe to clean the canal. Other suggestions included filing complaints, starting awareness campaigns (7.9%), starting an NGO (13, of whom seven were in Abkhas[83]), donating money or labor to cleanups, planting trees, teaching children the value of cleanliness, or simply avoiding pollution. The relatively frequent mention of awareness campaigns in the reflexive question should be noted: nearly 8% of those who would take action themselves mentioned it, whereas only 0.5% of those who thought people in general could do something mentioned it.

The most common environmental improvement activities attributed to the private sector by our respondents are planting trees, removing garbage, and establishing public gardens (64% of the positive answers); another 28% mentioned providing plastic sacks for garbage collection. Perception of the NGOs was even more lackluster: the most common answer was to lead awareness campaigns (30 of a total of 55 respondents, or 54.5%) or to help with garbage collection (20 of a total of 55 respondents, or 36.4%). Pretty much the same kind of answers, only fewer of them, were given in the case of political parties.

In general, these answers are consistent with the notion that the main concern of people is with the cleanliness of their immediate environment. The main focus of their own efforts should be on cleanliness, and what they expect from the government and from civil society organizations is in the same area. Since silences are sometimes interesting, one can note that no one thought of parties or NGOs as possible intermediaries in putting pressure on authorities to act.

Individual Solutions

People living in Kafr al-Elow react to the air pollution from cement dust by blocking off their windows and doors most of the time. They keep the windows shut, sometimes adding plastic sheeting, or cover the cracks with wet cloths to prevent cement dust penetration. The result is that there is no ventilation inside the houses, and hence aggravated indoor air pollution. They also are bothered by mosquitoes and flies, and make liberal use of insecticides in these closed interior spaces, or use strong cleaning substances to control the insects. Some women in Kafr al-Elow reported spending as much as LE50 a month on pesticides,[84] and one said to el-Ramly, with perhaps some hyperbole, “My children and I cannot sleep at night because of the mosquitoes, I'd rather buy pesticides than food” (el-Ramly 1996:86).

Housewives in Kafr al-Elow complain about the difficulty of drying their laundry in the open air, and they also clean daily to remove the accumulation of dust. If they cover their laundry with plastic sheets to protect it from cement dust, then the laundry takes much longer to dry. A few people in our sample areas, but more in middle-class Ma‘adi, use water filters or buy bottled water to protect themselves against what they see as the impure public water supply.

The Problem of Individualism and Lack of Cooperation

In Kafr al-Elow and Dar al-Salam, participants pointed out that people are too busy earning their living to devote much time to community action, and do not have a clear course of action. People also do not have the money to fund collective action. Many people cannot afford to pay the extra sums of money for garbage collection, though they have joined together to raise complaints to the local council concerning dumping garbage in the canal and emissions of cement dust. Moreover, many households rely on wage income from the cement plant, so are torn in their loyalties. In the end they are "passive and apathetic" (according to the focus group) and pursue only individual goals.

There is a lack of local leadership structures, say Dar al-Salam participants. People follow the example of their leaders, who are perceived as lax, uninterested, and individualistic. Turning the observation around, committed leaders would result in committed citizens. In the 1960s people thought they were building a nation, and accepted that larger goal, but this attitude no longer prevails. Sayyida Zeinab participants added that if people fail to act, it is not because of their lack of knowledge or awareness but because they feel the lack of law enforcement provides no supportive framework.

Abkhas participants noted that people today are too individualistic and do not work well together as a community. Intellectualism and television are perceived as part of the problem here. Even so, there is more cooperation than in the city. In the past people were more polite and more moral. Participants in both Sayyida Zeinab and Abkhas argued that only life-crisis rites bring people together. People only cooperate on social occasions such as death, illness, weddings, or birth.

Sayyida Zeinab participants noted that people nowadays are not as cooperative as they used to be, but they are still better in Sayyida Zeinab than in other places. This is because it is a *sha'bi* area, and relations between neighbors are seen as strong. On the other hand, another comment drawn from the senior women's focus group was that "in the past, people used to care for each other more than now. There was a sense of community spirit, people were like one big family. There are many people who have moved into the area and they are strangers. They have the money and with it, they feel, they can do lots of things. However, they lack the old principles and patterns of cooperation that prevailed among the old families."

Certain social divisions also undermine cooperation between citizens, particularly that between old residents and newcomers, which was mentioned in Kafr al-Elow, Dar al-Salam, and Sayyida Zeinab. (Abkhas has no newcomers.) In localities like Kafr al-Elow and Abkhas where there are big families, the rivalry among them is also a problem, and makes it hard for the village as a whole to cooperate. Kafr al-Elow participants say that local organizations (such as a CDA) would be divided by the rivalries of the big families so that cooperation is difficult. Also in Kafr al-Elow people say the distinction between factory workers and others hampers cooperation.

In a similar vein, in Dar al-Salam, one senior men's focus group participant commented:

> "Now there are many people residing in Dar al-Salam. They do not know each other, they are strangers to the area and they have come from everywhere. Now you don't know where your neighbors are from. It used to be very easy to identify who was working against the community's interest and who was offending neighbors. Now this is not possible any more."

Who are the Leaders?

Participants in each locality had theories about the kind of people (age, gender, length of residence) who should become community leaders. In our two more village-like localities (Kafr al-Elow and Abkhas) big families often took the lead, though some pointed out that this advantage disappeared when the families became rivals.

The older men in Sayyida Zeinab felt that the younger men should take the initiative. "The old generation lived through that [period when the Ministry of Social Affairs was active] and learned how to cooperate. Now they are old, they are still responsible, but the main movers and the initiative takers should be the youth." The young women thought that women's role consisted of informing their husbands about problems, and otherwise of keeping their house clean. These city women felt that rural women were more "free and flexible" when it came to action. Former residents of Sayyida Zeinab felt that in the old areas, the problem is that the old generation dominates the scene, while the young have married and moved out.

Abkhas participants pointed out that the generation gap and lack of resources complicate cooperation. Sometimes the energy of the youth is controlled by the seniors and focused by their concerns. Nevertheless, a group of young Abkhas women identified the young men as the group that can initiate activities as "they are energetic, mobile, and full of new ideas. However, they have to fight their way through traditional structures and the older generations who have a paternalistic attitude toward them." Men said that the village elders used the youth as tools. The young men see themselves as not being integrated enough in the village, they have occupational and educational concerns, and are often not living in the village due to study or work. But if they fail to take initiatives, then it is not clear that anyone else will. "The end result is lack of initiative on the part of the village community as a whole." One example of a youth initiative was to hire a tractor and bulldozer to clean out the garbage thrown into a dump next to the canal, but people soon resumed dumping garbage there so the benefit was short-lived.

Young women in Dar al-Salam and in Kafr al-Elow did not see any role for themselves, given their age and gender. In Dar al-Salam, they indicated that only older women could play a public role. Meanwhile, in Kafr al-Elow, young women said that the community clean-up had to be the work of young men, given the restrictions on what is considered proper for girls in a conservative community. The young girls of Abkhas, on the other hand, agreed that they had a role in health education.

Collecting Money for a Project

There are examples in all areas of individuals or small groups starting the process of collecting money from neighbors to accomplish a project of collective benefit, such as linking up to the sewage system, arranging for garbage collection, or planting trees—or even such activities as the construction of a school or mosque. We call this contingent social organization in the sense that it is contingent upon a particular need and disappears once the task is done. No permanent organizational structure is created, though the cultural model endures (Hopkins et al. 1988:120).

There are other examples of such contingent social organization in Egypt: the ubiquitous "rotating credit clubs" or savings associations (van Nieuwenhuijze et al. 1985:47–50; Wikan 1995:640), and the use of mediators to settle disputes (Nielsen 1998; Zayed 1998). This contingent social organization is both based on cultural understandings and contributes to gradual changes in cultural patterns. Whether collecting money for a joint goal or mediating a dispute, these institutional patterns are well known even though they do not give rise to enduring institutions. They are what Putnam (1995:67) called "social capital"—"features of social organization such as networks, norms, and social trust that facilitate coordination and cooperation for mutual benefit"—or what Unni Wikan (1995:637), writing of Cairo, named "cultural competence," defined as "the arsenal of knowledge, skills, attitudes, and institutional practices that people possess and can employ in coping." We can combine these two related notions into one called "cultural capital" and suggest that these skills and attitudes form part of the social context for action, along with the framing of the situation in cultural terms and, on the social dimension, power relations, especially with respect to national-level institutions and their personnel. Cultural capital is also part of the cultural construction of social life in that it represents knowledge and skills that are held by individuals but shared in broad outline with those around.

In Dar al-Salam and elsewhere neighbors joined together to collect money to bring in a private company to collect garbage, but the company is not reliable or regular in its services. Groups of local residents have also banded together to provide lighting for the streets, after several incidents. People also contributed money to solve the rodent problem that the residents were facing. After the sewage construction work in Dar al-Salam, residents in some areas hired a bulldozer to level the street.

At one point a group of young men in Kafr al-Elow tried to start their own system of garbage collection from the homes. They charged a modest fee. They distributed plastic bags to residents with the idea that they would collect them regularly from the homes. But the project failed because they were not reliable in collecting the garbage. Also some people were reluctant to let young men enter [the female domain of] their homes: "This is a traditional village community and not a city."[85] Another time, the local council organized garbage collection, for a monthly fee of LE2, but the collector here was also unreliable. Some felt that even this monthly fee was more than the poor could afford. The government also placed garbage bins in sites where people could dump their own garbage. But they were placed only along the Corniche road, away from the main settlement.

Dredging the Khashab canal in Kafr al-Elow sometimes creates problems because of the garbage it contains. A woman from Kafr al-Elow told el-Ramly (1996:70),

> We have paid the operator of the *kassaha* (loader) LE20 several times in order not to unload the garbage in the canal onto the bank facing our house. We have cleaned it several times, but people, especially the tenants, keep dumping their garbage there. This time we are determined to watch them and forbid them from doing that so that the trees don't die. We have paid LE30 for those three trees.

Another time, a group of people decided to remove the dredgings from the banks of the canal themselves rather than wait for the government, but this effort was not very successful. Many people have thought of filling in the Khashab canal, and replacing it with water pipes to supply water for irrigation and other purposes. In fact, a leader of one of the big families in the community formed a delegation to the municipality to request this, but without result. The canal remains open and continues to serve as a garbage dump perhaps because it still also supplies irrigation water to Kafr al-Elow's remaining farm land.

Eman el-Ramly (1996: 65) reports some cases of street cooperation among women in Manshiyet Sadat (Dar al-Salam).

> When the sewage floods in Manshiyet al-Sadat, one of the street residents collects money from the rest of the residents and hires a sewage worker to clear the blockage causing the flooding. The worker brings his equipment with him, typically long metal sticks, and charges LE10–15. Most residents cooperate willingly, except a few. The person who collects the money could be a man or a woman, but in most cases a woman, because women spend more time at home and are always there when the leakage takes place. Usually, ground floor flats suffer more from sewage problems, especially if the buildings are below street level. Respondents attribute sewage blockages to objects thrown in the system by careless women.

Women from the same streets told el-Ramly that the same procedure works when a dump has to be cleaned up: they join together to hire workers to clean it out, but people soon start dumping there again (1996: 67).

In Abkhas a local woman donated the land to build a new school. The woman donor contributed half the cost of the land, LE8,000, while each household contributed the rest at LE20 each. The school was built by the government. This combination of local contribution and government action seems to be common. On the other hand, some commented that this style of action is usually the result of one or several people who take the initiative and solicit contributions, and who do not always deliver, rather than the result of a collective decision to act.

A collective act by some Abkhas women was to build a communal oven. Thus, each woman keeps a clean house, since baking takes place outside the home. According to a focus group participant, "Keeping the oven clean is the responsibility of the users; women tend to enforce their own rules as to how to keep the place clean, and punish those who do not abide by the rules." The oven was felt to be more urgent than garbage collection.

Kafr al-Elow people work collectively on social occasions: for instance, to clean up the community for Ramadan (sweep streets, decorate them with Ramadan lanterns, etc.), or for weddings and funerals. Actually this task was done by the young men. Women commented that they knew that working together could help solve problems, but that it was hard to bring people together. Since the task was public, it fell to the young men.

In their desire to connect to the sewage system, people in both Kafr al-Elow and Dar al-Salam pooled their money and hired a contractor. But in many cases the person was not skilled enough for the job, and so the connections have many defects.

Individual initiatives, sometimes involving enlisting help from others, are also mentioned in Sayyida Zeinab. Many of these actions have to do with efforts to start small gardens, to plant trees, or to clean the streets, and most were unsuccessful in the end. Some of these projects were not supported by neighbors, while others were thwarted by the government.

One member of the Sayyida Zeinab senior men's focus group recounted how he himself had cleaned one area and planted trees and the like. However, neighbors started teasing him, and making fun of him. Additionally, his assistant, who was responsible for collecting money from the residents to fund the garden project, used to pocket some of the money raised. In the end, this man decided to break the water pipe and uproot all the trees. Today this same plot has turned into a garbage dump. There were several similar stories. For instance, one of the young men recalled how a member of the community started raising funds to buy trees; he collected the money and started cleaning up the streets, but the seedlings never arrived and the project did not materialize. One of the senior men argued that another effort to turn a deserted area into a public garden was thwarted by local officials. Once the trees had been planted, the area was handed over to local council for maintenance, which marked the end of the project. In another case, the young women reported that it was the neglect of the surrounding community that caused the project to fail. One of the senior women recounted how in her building she decided to start cleaning the space in front of her apartment, and the neighbors joined in. However, this did not last long, and the building is once again dirty. However, this same woman was able to arrange with a garbage collector to clean the area in front of the building periodically, and with her neighbors to share in paying the garbage collector. The senior women in Sayyida Zeinab also gave an example of a successful effort by young men to create a public garden.

The young women in Sayyida Zeinab reported that private shop owners wanted to clean and pave the roads, but they did not complete the project, blaming the municipal council for refusing to help. One of the young women, however, argued that they gave up of their own accord ("they became lazy"). The young men reported an effort by a resident to get the people of each area to clean up their own streets, but people lacked the motivation to continue, and the effort did not endure. Once, when a building collapsed, the whole community volunteered to pay compensation to its inhabitants through the mosque. The mosque became a center of relief, aid, and shelter to any needy individual. Similarly, there were donations after the 1992 earthquake. In fact, after the earthquake, people pitched in to repair a house that was about to collapse.

These stories illustrate the continuing efforts to improve one's neighborhood, even under discouraging conditions. The difficulty people face in some of these cooperative actions is to figure out how to take the first step, yet the stories also show some difficulty in carrying through an action, and the difficulties in persuading neighbors to join in and stay interested.

Dealing with Neighbors

People say they do not talk to each other a lot about environmental issues, which is consistent with the fact that they report spending most of their free time watching television rather than socializing. Nevertheless, the frequency of certain answers suggests that people may converse more than they admit.

Neighbors sometimes engage in antisocial or disruptive behavior. People are generally hesitant to put pressure directly on a neighbor. We asked people what they would do if neighbors were too noisy (n=2,307). Most would not react strongly: 37.9% would speak to them politely and try to reach an agreement, while 36.2% would do nothing, and 12.4% said they had no problems from their neighbors.[86] Altogether 23 individuals (1%) said they would quarrel, 3.1% would complain to the neighbor's relatives or to the *omda* or *shaikh al-balad*, and 6.7% would go to the police. Over two thirds (70.8% of 154) of those who said they would go to the police were from Abkhas, as were 97.2% of the 72 who mentioned complaining to relatives or to traditional authorities. The village appears more quarrelsome than the city. With this exception, the localities correspond approximately to the overall figures mentioned above. Women are more likely than men to find that they can do nothing about a noisy neighbor (41% to 31.7%); men are more likely to take action, so that, for instance, 39.7% of men, as against 36% of the women, would talk politely to such a neighbor and try to reach an agreement.

In Dar al-Salam social pressure was sometimes brought to bear on neighbors to get them to behave responsibly. Residents might pressure others to change: "Women who are more conscious as to the health hazards caused by garbage or sewage would enforce certain patterns of behavior [on others] to reduce the contact with known sources of pollution" (Senior Women, Dar al-Salam). Local people, supported by a representative from the Dar al-Salam

Local Council who was also a resident, pressured a man into leaving room next to a school he was building to serve as a passageway.

But more often residents in Dar al-Salam were reluctant to approach their neighbors. "Sometimes if you try to correct your neighbor, he gets angry, or at best ignores the request." So a strategy was to demonstrate good behavior in the hopes that others would imitate it. ("If I start first, then my neighbor will follow"—men and women both mentioned this though they did not all agree it would work.) They are even more hesitant to invoke governmental authority against an offender. People have complex relations with their neighbors, and filing a court case does not improve these. People do not want to put their neighbors at risk from the police (risk of a fine perhaps), nor to antagonize them. Only in one extreme case, where a woman thought that the behavior of her neighbor was encouraging rats and snakes, did she take the step of contacting the police. Another woman from the Dar al-Salam senior women's focus group wanted to complain to the police about a cesspit emptier who was charging too much and had a monopoly, but no one would support her. Looking back, she felt that the attitudes that prevailed were: "Fear, 'why should we interfere?' and 'who cares?'."

In Abkhas, participants felt that if the village is generally clean, this would encourage people to keep it that way. Paving roads and installing street lights would help. People should understand the different kinds of waste and the ways to dispose of each.

Another area where people's lack of consideration could harm others has to do with the proper maintenance of automobiles. So we inquired in the 1997 survey what should be done to make drivers tune up their cars so they pollute less (almost all drivers say they do this). Here the most common answer was nothing, since the drivers should be able to see the need for themselves (33.7% overall, equal by gender, most common in Dar al-Salam with 44.4% of respondents and least common in Abkhas with 11.9%). This is consistent with the general reluctance to interfere with others. The second most common answer was that people should complain to officials or go to court (18.2%), and another 15.1% said that this was the job of the traffic police. Only 4.1% would take direct action by talking to drivers and advising them. This group was most numerous in Kafr al-Elow (7.2%) and least numerous in Abkhas (1.2%). It included somewhat more women (4.8%) than men (3.5%). Some other answers were: to apply laws more strictly (2.9%), to switch to unleaded gas or install vertical exhaust pipes (3.3%), or not to ride with such drivers (1.3%).

Most of our respondents (88.6%) feel that people are less cooperative now than they were before (see Table 22). This corresponds to an often-expressed feeling in Egypt that the values of today have changed for the worse compared to the values of a previous generation. This widespread attitude means that people may not undertake initiatives that depend on the cooperation of others.

Complaining to Authority

People are generally uncertain where to go to seek help. They cite local government offices for the most part, but without excessive optimism. Where direct complaints seem unlikely to work, people seek out more powerful people who can then carry their complaint to the right ears.

Of our 1997 sample of 2,307, only 91 or 3.9% said that they had actually complained to a government official about pollution. The 91 included 56 men and 35 women; 65 were in Kafr al-Elow, 16 in Dar al-Salam, 10 in Sayyida Zeinab and none in Abkhas.[87] Of 95 complaints in Kafr al-Elow (there was more than one complaint per person), 52 (54.7%) were about cement, and 31 (32.6%) about garbage and sewage in the Nile (and canals). This last was the most frequent complaint in Dar al-Salam (12 out of 20 complaints, or 60%), whereas in Sayyida Zeinab the most frequent complaint was about noisy workshops. Men and women essentially complained about the same things.

Of the 91 complainants, only 16 thought their complaint had any result—in 12 of the 16 cases the result was that the complainant's home was hooked up to the sewage system. The other four appear to have been given a noncommittal answer about the future. Of the 16 successful complainants, 15 were men. The unsuccessful mostly blamed the responsible officials for inaction. Even the more active officials only came to see the problem, but then did nothing (14 cases).

Many people feel that more coverage by the media of their plight would encourage remedial action. One resident of Dar al-Salam wrote to newspapers such as *al-Gumhuriya* and *al-Mesa'* on behalf of a group to complain about the lack of schools in the area. He felt the complaints were effective in getting the government to respond. Kafr al-Elow people thought they could use the newspapers to complain against the cement dust emitted by the factories. People mentioned two newspapers that had given coverage: *al-Ahali* and *Fagr Helwan*.[88] Television is also desirable. There was once a TV program called *Agras al-khatar* (Warning Bells) which covered the pollution situation in Ma'asara.[89] People in Kafr al-Elow recount that a TV crew once came there to film, but found the atmosphere so polluted they refused to leave their vehicle and instead filmed through the window; not surprisingly no footage was aired. Yet so far nothing has come of this publicity.[90]

In the 1997 sample, 130 (5.6%) had heard of a factory or workshop that had been closed because of its pollution. This group was more or less equally distributed among the four sites, but men were more likely to have heard of this than women (7% of men, 4.3% of women). Most of the examples were of blacksmiths or mechanics' workshops, or a variety of other workshops. None seemed to be the case of a factory. Of the 130, four said they had played a role in the closing: three from Kafr al-Elow and one from Sayyida Zeinab, two men and two women. The two women complained to the district council in Kafr al-Elow, one of the men used the law to close a workshop in Kafr al-Elow and the other called the police in Sayyida Zeinab.

Perhaps because of the difficulty of direct action, when we asked what was the best way to deal with pollution in the respondent's area, the most common answer by far was to get the television to come and take pictures (922 out of 2,307, or 40%; 43% of women and 37.1% of men). Breaking down the answer by locality, it was given by 47.9% of those in Kafr al-Elow, 43.6% of those in Sayyida Zeinab, 41.6% of those in Dar al-Salam, and by 24.3% of those in Abkhas. Nevertheless, many answers indicated various other ways of complaining. Appeals to members of the district council were cited by 14.9%, and writing to the district council by 13.9%.[91] Those who mentioned knowing a member in the People's Assembly were 9.3%. The more aggressive solution of going to court was mentioned by 9.3% overall, including 25.7% of the respondents in Abkhas, and between 3.8% and 5.2% in the other localities. Solutions cited by a few included moving away (5 cases), transferring workshops and traffic from the area (8 cases), forming an organization to solve the problem (23 cases), finding an organization to prevent pollution (5 cases), and developing awareness campaigns (14 cases). Only 31 (1.3%) said that there was nothing that could be done; they were mostly in Kafr al-Elow (18 cases), and mostly men (22 cases). And another 39 (1.7%) said they did not know.

The notion of filing a complaint appeared again when we asked for suggestions of ways to make factories use filters. Nearly half our respondents (47.5%) mentioned sending complaints to the government and responsible people, or going to court, and another 8.5% would complain to the president, prime minister, environment minister, or parliament. Those who thought that nothing was possible were 22.4% overall, including 26% among men and 18.7% among women. Women were more numerous among the 9.3% who said they did not know what could be done. Those who would revolt, sit in, or tear down the factory were 2.8% overall, including 3.1% of men and 2.4% of women. The radical activists were fairly evenly distributed among the localities, peaking in Kafr al-Elow with 3.8%.

Interestingly, the concern on getting the television to broadcast pictures almost disappeared as an answer here (1% overall, slightly higher among men and in Kafr al-Elow).

Contacting Intermediaries

Another form of "complaining" is to contact a high-level person so that this person can use his influence on decision-makers. We had examples from each of our localities, and give some cases below.

Many respondents thought that it would be more effective to use a *wasta* (connection or intermediary) in combating pollution than to go to court (42.5% against 31.8% of all respondents), while another 6.7% mentioned soliciting television coverage, and 6.1% held to the notion of petitions and formal complaints to the district offices. Other answers were more widely split (awareness campaigns, work with the *imam* in the mosque, cooperation between people, government enforcement of its own laws, get people to be committed, appeal to members of parliament, etc.).

The people of Abkhas were the most committed to the idea of using the courts.[92] In Abkhas 67% of those who answered gave this response, compared to 23.5% for the idea of using a *wasta*, which was the preferred option in the other three localities. Women were more likely to prefer using an intermediary over appealing to the court (49.2% against 29.1% of all respondents) than men (36.2% against 34.3%).

Another recognized technique is to organize a delegation to call on officials. One person in Kafr al-Elow commented with regard to delegations: "We sent a delegation, we demonstrated, and we attacked as a result of what people suffer in this area. In other words, we protested peacefully, we begged, and we yelled, and the response was that, yes, we will do something. However, this was just a tranquilizer so we would leave them alone. Later the person in charge pretended that he heard nothing." Another person commented that a demonstration might work, but that he himself would not take the lead but only participate as a part of the crowd.

If people have steps they feel they can take to bring pressure on the authorities, those in power also have countermeasures. Industries bring pressure to bear on their workers not to protest against the pollution the industrial process creates. Management has convinced workers in the cement factory that their income depends on not using the filters. "I am a small employee and I want my bonus. The administration says that bonuses do not come except through this smoke and dust. Filters should be kept from functioning in order for production to increase. If production does not increase, we will not be able to live. I know that the cement dust harms me and my family, but the government says there is a cement shortage in Egypt. They will not decrease production for our sake. We are all poor, let us die, employees and Kafr al-Elow residents, but the production must increase." Thus, workers only agitate for their bonuses, not for cleaner production. "Why would we move for the problem of pollution as long as there is money, bonus, cash. God bless the chimneys, may God increase the production so the bonus will increase."

Petitions

Kafr al-Elow people recognize petitions as a protest technique but think that drafting and signing them is useless, since the responsible people already know of the problem and do nothing. Also collecting signatures can be dangerous, and demonstrations even more so. "No one has time for this because everyone is afraid of the government, especially these days, when they can arrest anyone." No one can collect signatures. According to one of the residents of Kafr al-Elow, "This act goes beyond my courage and ability. I cannot fight the government because it is the main source of the problem. It has all the solutions in its hands, and we cannot protest." The only example of a petition cited in Kafr al-Elow, mentioned above, was a petition to fill in the canal. Petitioning the government to fill in a canal is perceived as being less problematic than petitioning against a public-sector industry.

However, there are ways to mobilize effective pressure. Not only Kafr al-Elow but all the Helwan region pressured the cement company to utilize its filters. Several meetings to discuss the pollution problem were held in the local council, which experts from Helwan University attended. Residents felt that this collective pressure resulted in the hiring of the maintenance company to operate and maintain the filters. Since there continues to be doubt that the filters are actually used, this accomplishment must be considered precarious.

Voting and Elections

One can look at voting as an index of the ability of people to influence their lives through participation.[93] If government is to be held accountable, then voting in elections is an excellent way to achieve this—more peacefully than, say, revolt. The relatively low percentage of voters (19% of the sample overall) correlates with the absence of any mention in the surveys of voting as a strategy to follow to improve environmental conditions (though there was some mention in interviews and focus groups). Members of parliament are judged on their ability to provide services to their constituencies, and hence members of the governing party, who have better connections with the administration have a built-in advantage over opposition and independent candidates. Environmental matters do not yet, however, figure prominently among those services.

Candidates make many promises but then forget them.[94] People in Kafr al-Elow said, "Resorting to official government channels is not going to solve any of the current problems, because the government is giving a deaf ear to the demands and needs of the area." Government and elected officials will only act after people take the initial steps, and only if they are persistent. Solving a problem locally makes a lot of sense, for a common feeling is that "politicians never do anything." One evaluation of the situation is that "the government and the parties remember Kafr al-Elow only at the time of elections. At that time, the media present more about the problems, and the government is bound to address some of the issues raised." Another is that "we have no hope in anyone except God and the president of the country. No minister or member of parliament can do a thing."

In our 1997 sample (N=2,307), 635 individuals (27.5%) said they had a voting card showing they are registered as voters (see Table 44). The registered are 41.5% of the men and 12.9% of the women. The rate of registration is uneven between the different localities—68.6% are registered in Abkhas, 20% in Kafr al-Elow, 14.6% in Dar al-Salam, and 13.4% in Sayyida Zeinab. Consistent with other comments over the years (e.g., J. Abu-Lughod 1972), the village has by far the highest rate of registration, and also the most self-conscious strategy of using itself as a "vote bank." Most of those registered (81.9%) are registered in their home areas. Again there is some unevenness. The percentage of registrants who are registered where they currently live ranges from 97.7% in Abkhas to 90.8% in Kafr al-Elow, 75% in Sayyida Zeinab, and a low of 13.6% in Dar al-Salam. Thus Dar al-Salam stands out as having the largest proportion of voters registered away from their residence.

Among those in our sample who are registered, 69% actually voted in the 1995 parliamentary elections. The voters were 30% of all male respondents and 7.5% of all female respondents. Men were more likely than women to return to a previous residence to vote (18.1% of male registrants as opposed to 8.2% female). The highest rate of participation in these elections by locality was in Kafr al-Elow, with 80.8% of registrants in our sample voting, followed by Sayyida Zeinab with 71.3%, Abkhas with 66%, and Dar al-Salam with 62.5%. Of course the highest ratio between voters and the overall sample was in Abkhas, where the number of voters was highest. In fact, 52.3% of all voters we recorded were from Abkhas. Similar to the pattern of registration, 83.8% of the voters voted in their neighborhood. Here the greatest percentage who voted at home was in Abkhas (99.6% of those who voted), followed by Kafr al-Elow (89.7%), Sayyida Zeinab (73.7%) and finally Dar al-Salam (18.2%).

The two extremes with regard to voting are Abkhas and Dar al-Salam. Overall, Abkhas has the most voters, and the most people who vote at home, while Dar al-Salam has a very low percentage of registrants, and most of those are registered outside the neighborhood; not surprisingly it also has the lowest percentage of registrants actually voting.

In other words, Abkhas has 347 registered from the 506 inhabitants we interviewed. Of these, 339 were registered in Abkhas, and 229 (45.3%) voted in the 1995 parliamentary elections, all but one of them in Abkhas. In Dar al-Salam, with the lowest rate of participation, 88 out of 601 were registered to vote, only 12 of them in Dar al-Salam. In the 1995 elections 55 of these voted (9.2% of total sample), including 10 of the 12 registered in Dar al-Salam. Sayyida Zeinab is similar to Dar al-Salam except that more registrants were residents. Kafr al-Elow shares the low participation rate of the urban localities, but 90.8% of registrants are residents, and the rate of participation among registrants, at 80.8%, was the highest of the four localities. The overall picture thus supports the generalization that rural people vote more, since even the urban area with the highest percentage of registrants voting is the most rural of the Cairo localities. It also had a hotly contested election (Longuenesse 1997).

Table 44: Voting Behavior

	Total	*KE*	*DS*	*SZ*	*AB*
Total sample	2,307	601	601	599	506
Registered to vote	635	120	88	80	347
Voted in '95 elections	438	97	55	57	229
Voters as percent of sample	19.0%	16.1%	9.2%	9.5%	45.3%
Registered locally	520	109	12	60	339
Voted locally	367	87	10	42	228
Local voters as percent of sample	15.9%	14.5%	1.7%	7.0%	45.1%

KE=Kafr al-Elow, DS=Dar al-Salam, SZ=Sayyida Zeinab, AB=Abkhas.

Source: 1997 survey.

Elections in our two village-like localities, Kafr al-Elow and especially Abkhas, are more personal and more reflective of social organization than in the two more urbanized localities. Also both registration and turnout are higher. They illustrate some of the forms of social pressure that can be exercised through elections.

The most contested constituency in our sample in the 1995 elections was the one that included Kafr al-Elow. This is Constituency 25 of Cairo governorate (see Longuenesse 1997). Like all constituencies in Egypt, it returns two members, one of whom must be a worker or peasant. The opposition parties, especially the Socialist Labor Party, are fairly strong in this area.

In the 1990 elections, a factory worker (Muhammad Mustafa) was elected as an independent but then affiliated with the NDP. Workers in the iron and steel plant were able to work through him to get that factory to install filters to protect the health of workers. Despite this, in the 1995 elections Mustafa was defeated by another factory worker (Ali Fath al-Bab), also running as an independent, but with Islamist leanings.

The member of parliament holding the open seat from the constituency throughout the study was Dr. Muhammad Ali Mahgoub, from nearby Helwan al-Balad. He was Minister of Religious Endowments before the 1995 elections. He used his power as minister to authorize the building of a school (with funds raised locally) and to cancel the substantial debt of a club (the Kafr al-Elow Sporting Club, affiliated to the Ministry of Youth and Sports) rather than to support environmental issues such as reducing air pollution. Dr. Mahgoub is also said to have promised before the 1995 election that the mounds of dredgings would be removed from the banks of the canal, and that the filters on the cement plant would be installed and used, but none of this happened. "We protested to Dr. Mahgoub, and we said that every parliamentary representative had promised us before the elections to install the filters, and as the elections end their promises fade. Dr. Mahgoub said that the filters were there and had been installed, both the first and the second stage, and that they would function by the beginning of 1996, but nothing happened. Dr. Mahgoub has not been to Kafr al-Elow since election time [about five months earlier]. We wrote many protests to the newspaper."

Mahgoub's opponent in 1995 was Mustafa Bakri, a journalist and candidate of the Liberal Party. Neither candidate made much use of environmental themes. Mahgoub was narrowly reelected in the second round, then lost his portfolio and some of his power to deliver services. His 1995 campaign in Kafr al-Elow worked through the influence of large families to which he was allied by marriage; apparently most of his voters came from this network (Longuenesse 1997:262–63).[95]

The social organization of Kafr al-Elow is dominated by half a dozen large families (Fakhouri 1987:12, 55ff). These families have systems of support and solidarity, systems of punishment, and a hierarchy. The heads of families are strong enough to settle disputes among people, and accept a certain amount of responsibility for people. The respect due to family members who have acquired an education, especially the university graduates, is a new factor (p. 63). The big families have guest houses where marriages are celebrated and condolence ceremonies are carried out. When the Salloum family guest house was damaged in the 1992 earthquake, they collected a total of LE 65,000 and built a two-story building, with the ground floor for a Quranic school and the top floor for family gatherings such as weddings and funerals. Two families run charitable organizations that provide low-cost medical care and also sometimes extra lessons for students. These are financed by alms (*zakat*) offered by family members; sometimes the doctors work out of a sense of charity: "The poor are my relatives and I have to pay them back" noted one. "Foreigners," or outsiders, are not organized in Kafr al-Elow; the big families are strong enough to prevent outsiders from having their own organizations as they do elsewhere in the Helwan area. Instead, they are the tenants of the big families, who are then responsible for their behavior.

In the run-up to the November 1995 parliamentary elections, the villagers of Abkhas met to decide on a joint strategy to bargain for their collective vote. The willingness of the candidate to contribute to the village was a factor, as was, to some extent, natural sympathy for the candidate. Abkhas has voted for non-NDP candidates in the past, and this has caused trouble. At first the village tended towards a Wafdist. But people schemed to get a large payment from one or more candidates for the good of the village. The usual tactic at election time is to devise a village project to which candidates could make donations. So in 1995 candidates made a contribution to the cost of the land for the projects (a youth center and a CDA). Most of the village projects, such as having an independent agricultural cooperative or a youth center, were not environmental in nature.

In 1995, an informal village leader was put in charge of the process. "Village meetings were held to discuss the issue of taking money from the candidates, consent was secured, projects to be undertaken were agreed upon, and the distribution of votes among the candidates was discussed." In general, the feeling is that once elected, a candidate pays no further attention to them—hence the idea of getting the benefit first. The parties themselves play no particular role. A group of young women noted that "opposition parties can be useful only during election times, especially if one can manipulate the situation in such a way as to obtain the maximum benefit from political competition."

Only at election time is there any response at all to people's concerns, but election promises are often not kept. "When there are candidates who already hold important posts [in the government], there is very little that can be done not to have them reelected." Only at election time is there any sign of the opposition parties. The exception, noted by the Sayyida Zeinab young men's focus group, is the time when the Labor Party, under Seif al-Islam al-Banna in the late 1980s, started a street cleaning movement in Sayyida Zeinab, combined with garbage collection and some educational campaigns. People do not like to complain to the government about each other, but can speak and work directly with one another or seek local mediation. In some cases, they may refrain from action against someone for fear of harming that person.

On another occasion, Abkhas had to choose a new *omda*. One of the candidates promised that if chosen he would embark on various projects to clean up and beautify the village, applying to the national social fund for the money to do so. One specific item on his agenda was the purchase and installation of garbage containers. In the end, however, another candidate, not resident in Abkhas, was chosen, while this first man remained an influential but informal local leader. What is arguably the most urgent environmental problem in Abkhas, the lack of clean water, did not enter into these discussions.

Lawyers and Scientists

> There are scientists who can propose solutions to the problem of the cement dust, and there are religious figures who can approach the people from the standpoint of religion. Political figures can use religious idioms to convince the people to act in a certain way, and that applies to the question of the environment as well
>
> Citizen of Kafr al-Elow, 1996

Lawyers played a role in our three urban areas, taking action, sometimes confrontational and sometimes cooperative, about pollution issues. Several lawyers in the Helwan area (but not from Kafr al-Elow) have filed suits against the government and local public sector industries to force them to accept responsibility for the air and water pollution they create. These cases were undertaken at least in part to attract publicity, and regardless of their outcome they served to create a consciousness of environmental issues. The principal lawyer to file suits on environmental issues and pollution in Helwan is Mr. Muhammad al-Damatty, a lawyer in Helwan City. He filed several such suits in the late 1980s. The purpose, according to him, was to draw the attention of people to the pollution problem. The suits were to establish the facts of the case, using a court order to report the condition of the environment in Helwan. This is labeled a "status authenticity case." The case was filed in Southern Cairo Court in 1988, and resulted in an expert opinion in 1990 that the factories in question caused air pollution above accepted norms. The case then moved to the appeals court. "In fact all the orders were in my favor, and on the basis of this I can file compensation suits for the damages caused by air pollution. Since neither the government nor the companies showed any interest in this problem, I filed these suits." Several other people joined him in these suits—lawyers from Helwan, colleagues from the Tajammu' Party, and some journalists. Raising money to fund the case was a problem. There were also confrontations with the police, and other problems: "My car was vandalized and I have been threatened." Many of the copetitioners were intimidated and dropped out of the suit. The judges themselves were indifferent. "They consider these cases either an effort on our part to gain fame, or a threat to Egypt's reputation and a way to distort its image."[96]

Another lawyer filed a complaint against the government alleging that Dar al-Salam water was polluted. This complaint argued that the water from the taps is dirty, containing microbes, dust, viruses, salts, and bacteria, and stated that this was confirmed by several reports from health clinics. He argued that the excess of salts caused him to have kidney stones. He asked in his suit that a group of experts test the water to confirm its composition, and whether it could cause stones.[97]

In Sayyida Zeinab, lawyers working with the NDP established an "environmental committee" for the Sayyida Zeinab constituency. This committee undertakes research to identify problems in a scientific way and to find solutions. Also, some young people in the district founded a related association called the Association of Environmental Protection and Resources, again, with links to the NDP. The Association's goals are to create awareness about the environment and to protect natural resources in the area, to help beautify the area by cleaning the streets and planting trees, and to carry out research. Some local people remarked, however, that the political parties in earlier days, such as the Socialist Union of the 1970s, were more active and that people were able to accomplish their goals through them. At present the critical time to complain is before an election. For instance, before the 1995 parliamentary election, a resident complained about poor lighting in the street, and the candidate who was running for reelection hastened to solve the problem.

In 1994, Mona Ahmed Gad, a lawyer in Sayyida Zeinab associated with the NDP, drafted a well thought-out memorandum on environmental pollution in Sayyida Zeinab and suggested ways to resolve the problems (Gad 1998). She discussed air and water pollution, the problem of contaminated food, problems of garbage and noise, and smoking. She focused on the use of *mazout* (a kind of low-grade kerosene/fuel oil) by factories and ovens in Sayyida Zeinab, and also stressed pollution due to truck exhaust and public transport. The problem with water in the area was that rooftop storage tanks were not cleaned, and the problem with food was the use of pesticides by vendors to preserve their stock. She also pointed out the absence of public toilets, especially near a major bus depot at Abu Rish, and cited hazardous working conditions in workshops.

Gad's main argument is that all this pollution has an effect on people's health, and causes them to lose work days and also to spend money they can ill afford. Thus, she says, "the country loses thousands of production hours which affects the national income," and continues; "This makes the economic costs of being cured of illnesses caused by pollution very frightening, especially in a country like Egypt, which needs more production and more national income, with a decrease in expenses to the huge population." She further argues that her point is not to cast blame, but to encourage everyone (government, factory owners, workers, residents, and political parties) to take on their share of the responsibility for protecting the environment. All these "must cooperate to save the area of Sayyida Zeinab from being assassinated by pollution."

A prominent citizen of Kafr al-Elow is Dr. Fawzy 'Abd al-Samad, who is professor of soil science in the Soil, Water, and Environment Research Unit of the Agricultural Research Center at the Ministry of Agriculture in Giza. He is active in the Kafr al-Elow CDA, and promotes environmental awareness. In Kafr al-Elow he is considered a "man of culture" and a "scientist who knows what he is saying." 'Abd al-Samad also ran for parliament in 1995, affiliated to the Green Party, and received about 4% of the vote in the constituency (Longuenesse 1997:261). However, he told a researcher that protecting the environment was mostly a matter of protecting life and health, and that the movement best able to defend the environment was the NDP because of its links with power (Boutet 1997:156, 158).

Thus, in our localities there are individuals who promote environmental positions through legal action, party membership, and involvement in local social action.

Conclusion

People tend first of all to find that other people, like themselves, are the cause of pollution. This reflects their sense that the discarding of domestic waste in the streets and other public places is the main problem. However, they do not see themselves as very well equipped to deal with other people, although certain patterns for cooperation among people for short-term goals are well known and often used. There is a reluctance to approach offenders directly, and in general our respondents feel that others are not as aware of pollution issues as they themselves are. Hence, we often sensed that people would prefer that the government take a directive role in dealing with pollution, for example, by laying down rules and enforcing them strictly. In general, people find that the government is at fault for not coping with the existing pollution, but they do not blame it as a cause of pollution, even though one of the main sources people recognize is public sector industries such as cement plants.

As for other organizations that might help deal with pollution, they are barely on the map. People are not familiar with NGOs, few of which are present in the research areas, and they are not much more familiar with political parties or private industries. The role of lawyers in many of our localities should be noted; they file suits against the government and other polluters, although usually without much success. At the time of our research, the EEAA was just getting started and was not well known. It has since been incorporated within the new Ministry for Environmental Affairs.

People cooperate with neighbors to solve certain problems (like sewer extensions), complain to government offices (usually local government since it is more approachable), contact key individuals to get them to bring their influence to bear on the decision-makers, and perhaps use their votes. Direct political action seems unthinkable, and environmental issues do not play a major role in elections. Nevertheless, the idea is there, and reinforces the sense that those elected should provide services for those who elected them.

8

Sustainability and a Cultural Model of the Environment in Egypt

> Maybe we pollute, or maybe not; we are only working. What can we do? Who is the customer who will come to us at al-Qattamiyya or some other place? We have been working here for 20 years, as long as the people who live here. Besides, car paint and thinner have a good smell and no bad side effects.[98]
>
> Worker in an automobile paint shop,
> Dar al-Salam, 1995

Egyptians aspire to live in a clean environment, with paved streets, clean air, good water, reliable garbage collection and sewage disposal, and space and quiet. What does this aspiration mean, and what can ordinary people, or organized society, do to realize it? What kind of a cultural model of the environment are we talking about?

One misunderstanding that should have been scotched by now is that Egyptians do not care about their environment, or that they do not notice the difference between a clean environment and a polluted one. It is the contrary that is true. People are aware of most aspects of the problems that most immediately affect them, but not of circumstances beyond their experience. They are concerned with living day to day, not with overall sustainability. In general, our information is a reminder that people aspire to be stakeholders in the system, and that they have multiple opinions and attitudes, as well as certain kinds of knowledge. It is not so much that they are potential actors; they are already actors in this game of social change.

Our study bears on other issues that are central to current discussions in Egypt. As Gomaa (1997: 5–6) has pointed out, the domain of the environment is conducive for examining the relations between international concerns, governmental decision-making, and citizen participation. It also evokes the issue of the role of scientific knowledge in a social context, of how science relates to policy. We show the limited role of NGOs and grassroots actions, the impact on people of development programs and trends, the creative ability of people to organize themselves through cooperation and collective action, the patterns of governance, and the role of the state—in a word, the prospects for environmental activism and popular participation in public affairs. In addition, our study throws light on such issues as the place of gender and class distinctions in Egyptian life, and the differences and similarities between urban and rural cultures.

Although our study was not formally constructed to be "representative" of Egypt, we believe that it covers a major part of the situation, at least as far as the working and lower middle classes are concerned. We do not deal with the issue of the cultural constructions of the upper classes or the decision-makers, nor with their actions.

The Study

In this study we have combined several streams of data—from surveys, from participant observation, from interviews and focus groups, from the measurements of actual conditions in our research localities, all integrated into a comparative framework based on the social science literature. Our approach as a team was to concentrate our research on four localities in order to focus our efforts and to gain synergy from carrying out several types of research in a circumscribed area, so that the parts would be more clearly supportive of one another. The focus in our research, reflecting the preoccupations of those we talked to, was on the so-called "brown agenda" of cleaning up a polluted environment rather than the "green agenda" of conservation of nature.

We present our data from two points of view: in terms of social science methodology and in terms of how it may be used in programs of various kinds, notably reducing pollution in Egypt and perhaps elsewhere. Conceptually, we are presenting a cultural model of environment and pollution in Egypt, and we are interested in the implications of our research for a general understanding of Egyptian society and culture, both urban and rural. From a programmatic point of view, the question is whether any intervention is necessary or desirable, and if so how to influence the communities characterized by the cultural model described here, or how to provide the development agencies concerned with these issues with reliable data.

Research Questions

In Chapter 2 we presented several research questions that not only guided our data collection but also the presentation of those data in this study. We wanted to know about the scope and nature of the environmental situation in the research sites, and how these situations are understood or constructed by those who live and work in the sites. We were also interested in knowing how these environmental understandings were affected by a range of background factors, such as place of residence, gender, age, class, or occupation, and in trying to gauge how the understandings were affected by the mass media, by economic considerations (including those linked to economic liberalization), or by direct observation. When we found relatively few differences between the various background factors, we concluded that we were dealing with a cultural model. Furthermore, we looked at the attribution of responsibility for environmental deterioration, the sense of risk or threat from environmental factors, and the possible rise of social movements and other social action around these concerns.

Now is the time to see whether the hypotheses were borne out or not, and what the answers to the research questions are. We have reviewed the data in the preceding chapters, and given some idea of the answers in detail. In this conclusion, we will summarize our answers.

The Environmental Situation

Industrial and work activities and residences are intermingled in all our localities. The industrial activities range from the major cement and metal-working factories near Kafr al-Elow to the small-scale workshops and medium-sized factories of Dar al-Salam and Sayyida Zeinab to the agriculture of Abkhas. Automotive traffic is particularly an issue in Dar al-Salam and Sayyida Zeinab, and along the Corniche road in Kafr al-Elow, but in all areas is essentially restricted to the main roads because of the narrowness and poor surfacing of the lanes.

Over the past generation, the Egyptian authorities have made considerable efforts to provide basic amenities to the urban and rural populations in Egypt, but there is still much to do. In our study sample, almost all the urban households were hooked up to water and electricity, and a growing majority to sewage systems. Transportation is acceptable, though favorable opinions were perhaps artificially high because of our selection of communities along the Metro line. Solid waste disposal is lagging behind. However, living conditions are often crowded and noisy. People particularly complain about the way others use public space, the streets, and any open areas.

The measurements we have carried out in the four localities concur with much of the local evaluation. Although we could not measure garbage and sewage problems, they are clearly visible everywhere. Air quality is sometimes problematic, especially in Cairo, and especially because of particulate matter, where all our measurements, like those of others, far exceed international norms. Lead levels in the air are also high, sometimes above international limits. Water quality is generally acceptable in the city but unreliable in the village site. Noise levels are high without overstepping the boundaries of internationally acceptable limits.

Egypt is currently at a stage where the solution to problems is scientifically rather straightforward and does not depend on debate about scientific imponderables. The issue in Egypt is clean air and water, for instance, not global warming, acid rain, or deforestation, all issues that have generated a lot of controversy in the West. Still there is the need to link scientific discourse with the knowledge of the people: people know more than experts usually give them credit for, but not as much as they might. At a certain point, scientific knowledge itself becomes debatable, so the question of education is more complicated than simply passing on accepted truths. Instead, there is a need to involve all stakeholders in a dialogue, to combine the scientifically necessary with the politically and socially feasible.

The Cultural Construction of the Environment

We have stressed throughout this study that the environment as a field of action is culturally constructed. People's shared understandings significantly shape the actions that they may take. We have examined the ways in which this cultural understanding may vary by social background variables such as place of residence, gender, income, or level of education.

The people in our sample are fully aware of the existence of problems of environmental pollution, as is shown by the readiness with which people answered our questions and identified their problems. However, their evaluation and identification of the problems differs from that of the experts. The experts are more concerned about lead than our respondents, since lead is not perceptible through human senses, though it is nonetheless harmful to the body. Moreover, the experts, and for that matter the upper classes, are concerned about issues that this study does not touch on, such as clean seas and coral reefs, the relation between environment and tourism, issues of biosafety, or the overall water budget of Egypt. It is not awareness *per se* that is the issue, but rather the kind of awareness (and its sources and consequences). This awareness includes a sense of living with risk, as people are increasingly apprehensive about the environmental threats to food, water, air, and so on, with which they routinely live (Beck 1992).

The notions of health and cleanliness are linked. The relatively high level of concern about pollution results primarily from its perceived threats to health. The understanding of pollution is thus linked to healthcare, and to perceptions of health problems, the body, and so on. People are more concerned about the health effects of air than of water or noise. At the same time, pollution is considered equivalent to dirt, and hence the metaphor of cleanliness is central to the understanding of environmental pollution in Egypt. People also extend the image of dirt to the moral sphere and include in their concept of moral pollution misbehavior of various kinds, linked to drugs, thugs, and sex. The notion of dirt is also used to distinguish a clean "us" or "me" from a dirty and inferior "them."

Egyptians in our study are more concerned about pollution than about the natural environment. Environment for our participants tends to signify the situation in which people live, urban or rural, congenial or unpleasant. Ideas of "nature" are present, but are dominated by concern about environmental threats. These threats, in approximately descending order of the importance that our respondents give to them, are: garbage and sewage, air pollution, water pollution, noise pollution, overcrowding, and vermin (rodents and insects, especially mosquitoes and flies). This general pattern is substantially the same in rural and urban Egypt. Respondents ranked environmental pollution equal to poverty and inflation (money issues) at the top of a list of concerns. They outranked family affairs, population increase, accidental death, war, and crime, in that order. This high ranking is almost certainly because of the idea that pollution can lead to health problems, and also (as mentioned in the earlier analysis) because our survey dealt specifically with pollution issues. The high level of concern is also shown by the result that over three quarters of our respondents agreed that people have a right to clean air and water. The use of a term like "rights" is relatively strong language.

Our respondents are also aware that factories and workshops are possible threats. The work environment, whether in industry or agriculture, can be harmful both to workers and to nearby residents. Here the paradigm is the cement factories, with their highly visible cement dust, although some people also complain about factory wastewater entering the Nile. Although people know that factories contribute to a polluted environment, on the whole, since industry symbolizes development and employment, they would opt to retain factories with safeguards to maintain a clean environment as well, rather than do away with them in their neighborhood entirely. In rural areas, people recognize the pervasiveness of agricultural chemicals as a problem, but invoke a Faustian bargain in which the chemicals are necessary for a profitable agriculture.

The water issue dominates the rural areas. Rural people differentiate between different kinds of water, which they tend to use in different ways. Villagers believe, contrary to scientific analysis, that flowing water is clean and that produce taken directly from the fields is clean. Since flowing water is considered to be pure, washing and to some extent dumping in the canals is considered relatively acceptable. Water pollution is nonetheless understood mainly as a matter of dumping in the slow-flowing sections of the river and canals. The general awareness of water as a problem is not as sharp as the problem itself. This may be just as well, since there is no good source of water currently available to the villagers.

Effect of Independent Variables on the Data

Internal variation in our data was sharpest by locality. The most significant differences were between one neighborhood with its specific environmental problems and another. The next most important independent variable was gender, but differences were not as prominent as with locality. The linked factors of education and occupation played a small role, but bigger than that of age, which was not consistent. Class sometimes played an independent role, but not always, and sometimes its role had to be inferred from the differences between the localities, or from attitudes considering certain others as dirty or poor. We conclude that attitudes and understandings of environmental issues are primarily generated in and reflect the residential situation, in other words the neighborhood or locality, with its particular combination of threats and advantages. On the other hand, we sense that the older social structure of Cairo, as an archipelago of social islands and gated neighborhoods, is ceding to a newer form of organization tied to television and other mass media, and tending to create a mass society in which locality might be less significant.

Locality

The greatest and most significant contrasts involving the independent variables are those that result when the localities are compared. On many questions there are major contrasts between our four research localities, while on others there is a striking uniformity. The main differences are in the area of the physical circumstances of the different localities—the cement plant in Kafr al-Elow, the sewage problems and overcrowding in Dar al-Salam, the noise, dust, and old buildings of Sayyida Zeinab, the water and agricultural chemical problems in Abkhas. There are also differences that reflect social features of the localities. There are many areas where the two localities with the relatively more rural social structure (Kafr al-Elow and Abkhas) can be contrasted with the other two. We have suggested that this difference could be taken as a proxy for class or income, although it may also directly reflect the retention of large family structures in Kafr al-Elow and Abkhas compared with greater individualization in Sayyida Zeinab and Dar al-Salam. On other issues, there is remarkable uniformity, for instance, on the question of the relation of religious teaching to environmental issues or on the belief that people are less cooperative now than formerly. Similarity between our cases suggests that people are sharing in the same general culture. Thus, the differences between localities can be construed as due to the actual problems confronted or to the individual social structures, while similarities refer to general patterns and trends in Egyptian culture.

Gender

There is some patterning of responses by gender, but on the whole the parallels are more striking than the contrasts.[99] There is great similarity in many answers, broadly speaking those based on observation in one's neighborhood, or those based on ideas that are widespread in Egyptian culture. For instance, there is little difference in the answers from men and women to such questions as, "Where do you wash your laundry?" and "Where do you throw your used water?" even though it is the women who do these tasks.[100] The ranking of issues by their importance is nearly the same for both women and men; so is the ranking of dangers to health from air, water, and noise. The identification of major sources of air pollution is about the same. Both have essentially the same estimation of the environmental concern of those around them.

Differences are slight. Men worry more about factories for air pollution and microphones for noise, while women worry about vehicles. Men worry more about factory waste as a water pollutant, while women worry more about people dumping. Women worry more about drinking water, but less about water pollution, and more about both air pollution and garbage. Women are somewhat more likely to think that the government has done nothing to protect and clean up the environment. If there is a theme in all this, and it is a weak one, it is that women see the question more in interpersonal terms, and men more in technological terms.

One gender difference is that men are more at ease in thinking about contacting the outside—public—world for help. Thus, for instance, men are more interested in the environment, and discuss it more with friends. Women are less likely to know organizations that deal with the environment, and less likely to think they know where to go with a problem; nearly half of the men identified local councils as an organization that deals with the environment, and a quarter said they would seek help from a government office of some kind. Women have fewer ideas of what to do to improve matters, but the ranking of choices is the same for both categories. Consistent with literacy rates, women are more likely to watch TV (and thus to have seen TV spots on the environment) while men are more likely to read newspapers. This is one of the biggest contrasts. However, men and women cite "mass media" in general as their main source of information in learning about the environment. Men are more likely to have heard of the laws relating to the environment. Finally, women are slightly more pessimistic about the environment, and by a small margin they would rather clean up the environment than build factories.

Consistent with figures from elsewhere in the world, women have a lower opinion of their health, and more specific complaints, than men. Thus, women are less likely than men to rate their health as excellent or good, and more likely to cite one of the complaints we asked about. However, the frequency of the particular complaints follows about the same pattern in both genders. Thus, for instance, headache pain is the complaint most commonly cited by both genders, followed by sleeplessness and nervousness.

Class

Occupation and education, as well as locality, are among the proxy variables for class in our study. But also some of the answers and constructions people offered to us reflect class positions. Although it is neither stressed nor virulent, class feeling, or resentment, can be inferred from the belief that the poor suffer more from pollution than the rich, and in the feeling that rich neighborhoods are favored by the government. People could but don't point to the role of civil servants in the upper echelons of the government bureaucracy who make the decisions concerning public industry and gasoline, and who drive cars that contribute to air and noise pollution. Feelings of environmental justice (or injustice) are present but weak. The class feeling of the rich toward the poor is seen in the argument that the poor are responsible for pollution because they are dirty and pollution is dirt, and in the insistence on education of the poor to the standard of the rich as the solution for pollution problems.

Construction of the Social Dimension of the Environment

The Media

People think of television and newspapers as a source for their ideas, but they also dream of them as a way to attract the attention of the upper classes to their situation in the hope of improving it. People say that their awareness, or knowledge, is largely drawn from the media (newspapers and television), though our data belie that. The sometimes rather substantial differences in knowledge and awareness between the localities reflect the actual situation in those localities, whereas people are essentially exposed to the same national newspapers and state television, in other words, the same stimulus. If the media are the same for everyone, they can not be used to explain differences.

Among the media, television certainly has the widest exposure, particularly among women, though nearly half of our sample read a newspaper every day. Our evidence is that the breadth of exposure to television is complemented by the depth of newspapers, so that newspaper readers are better informed. We did not tap into any debate about the role of science, and especially about the significance of competing scientific interpretations. Of course, another medium is talk among people, and that is also seen in our results, though respondents tend to downplay it.

Who is Responsible?

On the whole, Egyptians think that the source of environmental problems is other people, who should be cleaner, make less noise, and not smoke where it bothers others. The typical act of uncleanliness is when neighbors discard garbage (including sewage, and used or dirty water) in the streets. This act is something everyone can see, and probably most people have practiced, and so is understandable. However, this behavior is not thoughtless, as is shown by the self-consciousness with which people attempt to dissemble what they are doing. It partly reflects the absence of viable alternatives.

Our respondents are also aware of the responsibility of factories, workshops, and vehicular traffic for pollution, but on the whole (there were some exceptions in Kafr al-Elow) they did not place significant blame on the managers of industrial sites or on the upper classes in general for the decisions taken by the government. Although the private sector dominates the small industries and workshops (including some controversial lead smelters in residential areas), it was even less criticized. The major exception to this is the willingness to blame those responsible for not consistently using the filters on the cement factories, but this negative opinion is pretty much limited to that particular decision, and is not generalized to a category.

People generally have a feeling that there are threats in the environment, in particular to public cleanliness, but also to food quality, where they do not have the tools to make choices, as well as to general issues of air quality, overcrowdedness, and noise levels, where they basically feel overwhelmed and unable to affect the situation. They tend to focus their sense of what they can do on their immediate circumstances, which means resenting or trying to influence the behavior of their neighbors. From this point of view, the argument is in the form of an implicit syllogism: pollution is dirt, the poor are dirty, therefore the poor are the cause of pollution. This syllogism is certainly present among the nonpoor, and even in our sample, the richer neighborhoods found people they perceived as poor to inculpate. This feeling was even sharper in Ma'adi. The argument that the dirty poor are the main cause of pollution seems to be accepted by the poor as well, who argue that they cannot help their behavior because of the lack of alternatives—an argument that has a good deal of truth in it. Thus, insofar as threats in the environment (a sense of risk) leads to action, it is largely action directed by people toward themselves. When people envisage their own role, it is in terms of greater self-discipline in managing waste, rather than as a mobilizer or catalyst in a confrontational situation. This inward direction of action sometimes amounts to blaming the victim on the one hand, and self-blame by the victim on the other—since the poor are the major victims of the behavior that they deplore and stigmatize.

Frames and Action

There are various frames available to Egyptians seeking to attribute meaning to the environmental and social issues in cities and countryside, frames in which intentions to act can be formulated. These frames are in effect metaphors. They could be labeled "the good prince," the "ignorant citizen," the "egocentric citizen," "social or environmental justice," "threat to health (toxic risk)," "environmental degradation or loss." All of these, but especially the first two, are evident here. These frames are a response to environmental deterioration but they are also a product of Egyptian culture, and reflect more general frames for action on the Egyptian social scene.

According to the "good prince" frame, the rulers are just and caring, but information on the true situation is withheld from them by intermediaries, so that action is required to reach upwards beyond these barriers and call a catastrophic situation to the attention of the rulers. This notion fits with the idea that initiatives are the responsibility of the government. The appropriate actions are to publicize one's story through the mass media or to appeal to an intermediary. According to the "ignorant citizen" frame, the problem is that many individuals are careless or malicious in their individual behavior, from ignorance of the implications of their behavior, so that the action required is education, or perhaps force, to oblige people to conform—basically to be clean rather than dirty. Combined with this is the idea that knowledge leads directly to behavior. Thus, pollution is blamed on the dirty habits of the poor, whose behavior can be corrected through education. This frame fits top-down action: people should be educated to match the standards of the upper classes. But in Egypt this image of ignorance is sometimes accepted by the poor themselves, a version of "blaming the victim." A related frame is the "egocentric citizen" frame, which holds that cooperation is difficult because of the egocentrism of others. This frame accounts for the appeal to people of the idea of a police force that would intervene and enforce rules, much like cleanliness is enforced, they say, by fines on the Cairo Metro.

These three frames, in various combinations, predominate in Egypt. The notions of "environmental justice," "risk from toxic contamination," and "environmental sustainability," or "threat to the resource base" are subordinate. The "environmental justice" frame is present in the ideas that the poor suffer more from environmental degradation and that people have a right to a clean environment, but it is not developed. Actions according to these frames might well run afoul of power considerations in Egypt, notably the difficulties of organizing large-scale collective action either through organizations or demonstrations and other forms of public pressure. Certainly people are conscious of the limits power puts on their actions.

Action does not automatically follow from definition alone since a major reason for inaction has to do with class or power: people fail to act because they feel blocked, they would be repressed or ignored, and unable to influence the decisions they feel are essential. This can lead to a turning inward of the behavioral problem, so that people appear to blame themselves. The Egyptians in our study certainly accept the notion that the environmental issues of concern to them are the result of human actions, and that they are going to be solved, if at all, by other human actions. The unexpected element is that they feel that these remedial actions are their own. Thus, they often frame the issue not as persuading other people or the government to set things right, or as the need to direct their ire upward, but as correcting their own behavior in order to improve the situation.

Yet action there is. One line of argument is that Egyptians are largely motivated by family issues, by providing for and advancing their family (Singerman 1997). Another line identifies the many small actions that individuals take, what Bayat (1997) calls the quiet encroachment of the poor, rather than any large social movement. One can also find and analyze behavior related to religious norms of one kind or another. More concretely, there are certain forms of collective action—the savings associations, collecting money for a one-time improvement (sewage and electricity, but also parks and trees), trying to establish a regular system such as garbage collection—which we call contingent organizations, since they last only as long as the immediate task that inspired them. In addition, people form small groups (of up to half a dozen members, perhaps) to call on officials or intermediaries to get them to intervene and change an undesirable situation, from a sewage overflow or excessive smoke from burning garbage to a factory emitting waste. There is concern, and action, but to speak of activism in the sense of a social movement would be stretching the point.

Culture and Theory in Environmental Change

This study has illustrated the usefulness of an approach to environmental change in terms of cultural construction, the first of the four theoretical models we sketched out in the first chapter. Our emphasis has been not only on the observable reality of the situation but also, mainly, on what people think about it, and in particular, what we can determine to be ways of thinking that are social in their origins and cultural in their content. From a theoretical point of view, we consider this study to be a contribution to this line of thought.

The comparison with the U.S. situation, as analyzed by Kempton et al. is illustrative of this, as we have seen in Chapter 1. After making the useful point that environmental values are integrated with core cultural values, these authors summed up the U.S. situation as follows:

> Our informants see nature as a highly interdependent system in a balanced state, vulnerable to unpredictable "chain reactions" triggered by human disturbance. Global warming is understood using the prior cultural model of pollution, the ozone hole, and photosynthesis . . . . environmentalism has

> already become integrated with core American values such as parental responsibility, obligation to descendants, and traditional religious teachings . . . biocentric values—valuing nature for its own sake—are also important for many. (Kempton, Boster, and Hartley 1995: 214)

One might by contrast sum up the Egyptian model of environmentalism as follows:

> The environment contains many risks, largely unknowable to ordinary people. These risks are a threat to health. Nature is not the issue; rather health is. There are certain key areas of importance, mostly linked to the notion of dirt, but including moral pollution. The responsibility for these problems is mainly individual, and the remedy must start with individual and local cooperative action. However, the deterioration in the quality of interpersonal relations complicates this. Ideas about pollution link with values on human relations in family and neighborhood, and on cleanliness and moral behavior.

There are some interesting points of similarity between the Egyptian and the U.S. models. Both, for instance, placed economic factors second to long-term environmental protection. However, there is a significant point of difference. The key metaphor in the U.S. case is "nature"—human beings are seen in the context of nature, and environmentalism is seen as protecting nature. The focus on nature also links to certain understandings of religion. In the Egyptian case the key metaphors are "health" and "cleanliness" (and its opposite, dirt), and all that they imply. Here the link to religion is through cleanliness rather than through nature. People living in overcrowded urban conditions, or even rural conditions, are concerned with the immediate quality of their lives. The focus is on the relations among human beings, ranging from neighbors to top government officials, rather than on human links to nature or even natural resources.

Achieving Sustainability

An interesting framework for further investigation is the notion of the management of the environment as common property. The basic dilemma is that everyone has an interest in a clean environment, and yet many act counter to that goal. The shortcoming of the "common resource management" approach is that it focuses perhaps too much on local governance, the management of clean streets and quiet times among neighbors, and does not include, as it might, aspects of the commons that extend beyond the local community. Many of the environmental issues that Egyptians face arise at the regional or national level, outside the local community, and find their solution there as well. Such issues as the need for a national plan for garbage (solid waste) disposal, the elimination of lead and other chemicals from gasoline, or an industrial and environmental policy that severely reduces emissions from cement plants and other industries, all reach beyond the local community. Instead, they imply a national level of governance, and hence such issues as the accountability of elected and appointed officials.

The Egyptian material provides some data on the two solutions that Hardin proposed (collective management or coercion through democratic force), and also support the argument from many anthropologists that people can find a way in tune with the general cultural setting to manage their common property. What is missing in Egypt are accepted mechanisms to coerce the public authorities from below, in order to establish accountability. Egyptians sometimes talk as if they would like strong direction from a central government, ready to enforce rules with punishment if necessary. Whether they would like such strong direction if they saw it is another matter. But perhaps behind these sentiments is the idea that this would be acceptable if everyone, regardless of social status, was subject to the same rules.

And the system does change. Sewers are built, landfills are planned, lead is replaced in gasoline by MTBE, natural gas is promoted as an alternative fuel to gasoline. This is the realm of power, where ruling individuals make decisions that may be in the interest of the poor but are not the result of pressure from the poor. If anything, they are to forestall popular action in the form of revolts. Gomaa (1997: 5–6) argued that the elite was motivated by general sensitivity to external political issues (the political value in being up to date on the world scale in fighting pollution, or the value in being the leader in the Arab world) rather than by concern for the environment. The ruling elite members who are concerned about the environment sometimes have a different set of issues, ranging from protection of the quality of life in such upper-class enclaves as Ma'adi or Zamalek to coral reef protection to anxiety over the general water budget of Egypt as the downstream country in the Nile valley.

Leading Egyptians, such as the minister of state for environmental affairs, have articulated the need to ensure a clean environment for a larger number of Egyptians in the future. Egypt's environmental policies need to aim for sustainability not only with regard to natural resources like water, but also with regard to livability issues in the urban and rural human environments. Sustainable living means good health in a relatively pleasant and stress-free environment, in other words, it means a control on pollution. Ultimately these projects require the cooperation of the people. The "stakeholders" may be collaborating with projects initiated from above, or they may be cooperating with each other to improve the situation around them. Their input must be recognized and valued. And without this participation, these projects will fail. Our study has analyzed the attitudes, opinions, and experiences of many people in urban and rural Egypt. It provides a baseline of information for imagining forms of cooperation that can develop and protect an environmentally sustainable and exemplary Egypt in the 21st century.

God Bless the Chimneys

Egypt needs more cement,
says the government.
Ditto the workers.
'We know dust loves
our lungs, sticks to them
and gets attached,
we know our blood's poisoned
like the irrigation canal.
We're all poor
as the desert sand;
let's die, residents
of Kafr al-Elow.
Output must increase;
how else will there be
money, bonus, cash.
Why should we press on
to install the filters.
We must live.
God bless the chimneys.'

—Sharif Elmusa

Sharif Elmusa, a scholar and a poet, reacted to his reading of the manuscript of this book with this poem.

Notes

1 EEAA, "National Report on Environment and Development in Egypt," presented to the UN Conference on Environment and Development, 1992, p. 41 (quoted in Gomaa 1997:35).
2 MTBE, or methyl tertiary butyl ether, is also used as an additive to gasoline in the U.S., where there is currently concern that it may leak from storage tanks and contaminate water supplies.
3 This could be seen in the papers and exhibits at the "Environment '97" conference sponsored by the EEAA in Cairo in February 1997.
4 Material available in the literature would allow the definition of similar cultural models from other parts of the world, such as Sweden (Lindén 1997) or India (Baviskar 1995, Gadgil and Guha 1995, Sengupta 1999), but this is not attempted here.
5 "Since the middle of this century the social institutions of industrial society have been confronted with the historically unprecedented possibility of the destruction through decision-making of all life on this planet" (Beck 1992:101).
6 For further discussion, see also Buttel 1987, Edelstein 1988, Fitchen 1988, Petterson 1988, Wildavsky and Dake 1990, Kottak 1992:296.
7 "Whenever there is dirt there is system . . . . The idea of dirt takes us straight into the field of symbolism and promises a link-up with more obviously symbolic systems of purity" (Douglas 1970 [1966]:48). For more essays on this theme, see Douglas 1992.
8 Obviously, such an assumption raises a host of theoretical issues in the social sciences that cannot be dealt with in detail at this point. Let it simply be said that we are aware of the problematic nature of the issue. Ideas that people take to be real affect acts and outcomes. Fowlkes and Miller (1987:69), for instance, paraphrase W. I. Thomas's remark (1923) that "situations which are defined as real are real in their consequences."
9 A T-shirt slogan reads: "Denial is not a river in Egypt."
10 This might be taken as a description of current Egyptian government policy.
11 For case studies see Proceedings 1985, McCay and Acheson 1987, and Berkes 1989; for a theoretical analysis, see Ostrom 1990.
12 Environmental movements are placed among the "new social movements," along with gender, peace, and similar movements.
13 One might almost say, blame the victim, since the poor are often held responsible for their own afflictions. Members of the elite are often quick to blame the poor, arguing that their misery arises from their own behavior.

14 This account is based on the following newspaper articles: Abdel-Salam 1999, Awad 1999, Bakr 1999, *Egyptian Gazette* 1999, Khalil 1999, Montasser 1999, Tadros 1999a, 1999b.

15 There are parallels between this case and the uproar about water quality in Amman, Jordan, in the summer of 1998, which led to the downfall of the Jordanian cabinet (see *Egyptian Gazette*, 1998). Apparently, shortcomings in the water-treatment plant meant that upper class and middle class homes were supplied with visibly impure water. Since hypothetically some of the water came originally from Lake Tiberias, opponents of the government saw the opportunity to blame the Israelis and the Jordanian government that had dealt with them. More prosaically, water officials were scapegoated before being exonerated. (For background information, see Abu-Taleb and Salameh 1994.)

16 A Mustafa Hussein–Ahmed Ragheb cartoon in *al-Akhbar* (November 1, 1999) showed a solemn government official explaining to reporters: "And the people are accusing the government of not knowing the reason for the cloud, but we are proud to announce that using infra-red photography we were able to see a devil who has a sick stomach and a lot of wind, and here is a picture of him letting it out from the top of the highest building in Greater Cairo. May God help us."

17 Quoted in *al-Ahram*, October 31, 1999.

18 After a somewhat milder version of the black cloud reappeared in November 2000, people began to consider this an annual event (Nasr 2000; Kadri, 2000).

19 In the material that follows the following abbreviations for the four localities are used: Kafr al-Elow (KE), Dar al-Salam (DS), Sayyida Zeinab (SZ), and Abkhas (AB). The information on the four sites is presented from south to north (KE, then DS, followed by SZ and AB), except in clearly marked exceptions.

20 Sewage and solid waste are often dumped in the Khashab canal. The engineering team reported that "the Khashab canal is heavily polluted, with the prevailing anaerobic conditions contributing to foul odors and the generation of hydrogen disulfides and ammonia among other toxic and obnoxious gases, with poor health implications for the residents."

21 The crowding in Dar al-Salam means that it is one of the areas of Cairo sometimes given the epithet of "popular China." See Hoodfar (1998:39) for another.

22 A tank or *transh* is typically a cement cube or tank of about 2 m to 3 m in diameter, buried in the street near the building, and into which the building's waste water empties. When the tank is completely enclosed, it can be called *khazan*; when the bottom is open to allow seepage out, it can be called *biara* (Nadim et al. 1980:96). They must be periodically emptied.

23 The illiterate in the sample made up 27.1% of the total. This figure of 45% for newspaper readers thus corresponds approximately to the 48.5% who were educated to the preparatory level or above. In between are those who claim to read and write and those with only a primary school education (Table 1).
24 The EEAA gave the figure of 12,662 tons of municipal solid waste per day in 1997 (American Chamber of Commerce in Egypt 2000:6), quoting EEAA "Solid wastes in Egypt and a comprehensive plan for overcoming wastes," 1997, p. 14.
25 "Lead has long been recognized as a neurotoxin that causes renal damage, neurological dysfunction, anemia, and at high doses, death . . . scientific evidence showed that lead retarded the mental and physical development of children, causing reading and learning disabilities; changes in behavior, such as hyperactivity; reduced attention span; and hearing loss, even at low levels of exposure . . . .Several studies . . . have also related increased blood pressure and hypertension in adults to elevated blood lead levels, which was shown to increase the risk of cardiovascular disease . . ." (Lovei 1998:1).
26 We did not measure lead levels in blood. Studies before the reduction of lead in gasoline showed that the average lead level in the blood of Cairo men was 30 micrograms per deciliter, and slightly lower for women. The exposure came mostly from food and water, though much of the lead entered the environment from highly leaded gasoline and lead smelters (PRIDE 1994, vol. I, p. I.ii). The PRIDE report (1994, vol. II, p. A-19) notes that "About two thirds of lead emissions to the atmosphere in Cairo come from automobiles fueled with leaded gasoline, with the remainder from the dozen or so lead smelting operations in the city."
27 Our data can be compared to those given in the study on neighborhoods without regular water supply by A. Nadim et al. (1980), and by Tekçe et al. (1994), describing the community of Manshiyet Nasser in greater Cairo.
28 Our surveys were done before the spread of the cell phone.
29 Multiple answers were possible on this question.
30 Friends are cited by 19.3% of the illiterate and 23.2% of those with post-secondary education; relatives are cited by 31.7% of the illiterate and 25.9% of the post-secondary people.
31 This was partly to see if feelings of class were expressed in distinctions between rich and poor neighborhoods.
32 Al-Ma'asara is a poor neighborhood between Kafr al-Elow and Ma'adi. Al-Basatin, a former village, is a poor neighborhood next to the cemeteries east of Dar al-Salam and south of central Cairo. Bulaq is a lower-class, inner city area. Al-Madbah (the slaughterhouses) is in south-central Cairo, and is considered poor and rough.
33 On street food, see Nawal Nadim's study (N. Nadim 1985) of a community that included a group of families who earned a living selling *foul madamis*

in the streets of Cairo in the 1970s, and Loza (1992) on street food vendors in Minya.

34 A high official of the General Sanitation and Beautification Authority was quoted as saying that "the residents of popular neighborhoods refuse to pay LE2 per month to the governorate's garbage collectors. They prefer to throw their garbage in the street" (Abdel Salam 1997:4). This last sentence is representative of upper-class scorn for the poor. See also the same phrase in Baracat (1998:4), who concludes that people must be taught that letting garbage accumulate in the streets is hazardous; our study shows that people are well aware of this danger.

35 In Egypt as elsewhere what is thrown away varies by economic class. Discards in poor areas are already generally pretty thoroughly picked over for reusable items. One EEAA official concluded that "even [sic] the garbage of the poor is worthless," reflecting both class disdain and a certain realism (El-Cherbini 1997).

36 The information from the focus groups and other sources for Dar al-Salam suggest that garbage collection is pretty unreliable, so this high figure for garbage collection may be chimerical.

37 On the garbage collectors or *zabbalin*, see El-Hakim 1981, Meyer 1987, Assaad and Garas 1993, Volpi 1997, and Abdel-Motaal 1997. Every so often there are efforts to change this system.

38 This may be because children carry it to the dumpster and cannot manage to throw it in (Hopkins et al. 1995:10).

39 See also the account of emptying the sewage tanks in Nadim et al. (1980:103). Tekçe et al. report (1994:37) that in Manshiyet Nasser human waste is discarded in "private pits cut out of the limestone rock in front of or alongside each building. These are rock vaults, not soak pits, so gray water (waste water) is introduced sparingly. The pits are emptied periodically. In 1984 this was done either by tanker vehicles with a pumping capacity or by laborers who carried the waste out of pits using buckets hung on shoulder poles. In both procedures, the sludge was carried only short distances before being dumped into empty spaces." Manshiyet Nasser is largely built on rock, so the difficulties of creating a pit are greater. Eventually, by the late 1980s, many pits were connected to the sewage network, but the connections were made by the users, and were sometimes faulty, leading to leakage and spillage, and a serious risk of infection.

40 In the 1970s, evacuation was manual, and the wagons were pulled by donkeys (Nadim et al. 1980:102).

41 Female respondents gave this answer 91.2% of the time, while 82.3% of male respondents gave it. We have no explanation for this gender discrepancy.

42 El-Ramly reports that 9 of the 14 Ma'adi women she interviewed were dissatisfied with the quality of tap water. "Most respondents have filters attached

to their taps, used to have filters installed, or plan to use filters. Some of those who use filters complain that, even with the use of filters, the water still contains residues. A few others drink bottled water, and two used to boil tap water but stopped because boiling changes the taste of the water" (el-Ramly 1996:63).

43 See the detailed analysis of water sources and use in other Minufiyya villages in El-Katsha et al. (1989:71–73).

44 Current factory estimates of dust reduction are much higher.

45 The filters are mostly electrostatic precipitators and are extensive machines in themselves. At the Turah Portland Cement Company in Turah, between Ma'adi and Helwan, they are managed from a separate control room and can be turned on and off by a switch.

46 The cement factories now say they halt production when the filters need maintenance.

47 A conference paper at the *Environment '97* conference sponsored by the EEAA and the Technical Cooperation Office for the Environment detailed a variety of ways in which cement dust could be used. A question from the floor asked why the dust was not used in these ways if it could be, and the chair of the session ruled the question out of order on the grounds that it was a political question.

48 Engineers at the Turah Cement Company say that much of the dust trapped by the filters is recycled into production.

49 At the Turah Cement Company, the older kilns use the wet system, and the newer ones use the dry method, with a much higher level of production. The dry method was introduced there in 1982.

50 Ma'adi is the nearest elite residential area to Kafr al-Elow.

51 An item in the *al-Ahram Weekly* for March 12, 1998, reported that five new cement factories were to be built near inhabited areas outside the city of Beni Suef, about 100 km south of Helwan. One factory was reported to be planned for a site about 300m from an existing village. Many people had raised the question of locating these factories 30 km into the desert, as required by law, but it appeared likely that by the time the authorities (i.e., the Ministry of the Environment) act, construction would be so far advanced that relocation would be costly. So the dilemma faced by the people of Kafr al-Elow is likely to be recreated (M. Tadros 1998). On the Beni Suef situation, see also F.N. Ibrahim (1996:190).

52 Most of our answers came from the three urban sites, since only 29 individuals from Abkhas answered this question, of whom 17 and nine cited other people, respectively, as sources of noise and traffic. In her interviews with women in Manshiyet al-Sadat (Dar al-Salam), Kafr al-Elow, and Ma'adi, el-Ramly found that women in Kafr al-Elow were not bothered by noise, that women in Ma'adi were so bothered, and that women in Manshiyet al-Sadat were divided over the same issue.

53 A newspaper article analyzing the phone calls made to Cairo's emergency number reports that about 25% were complaints about noise, while slightly more than 25% involved problems with water, electricity, or sewage. Those who staff this emergency number noted that "those who complain about noise were usually reluctant to give information about themselves, as they are usually complaining about their neighbors." They believe that complaints about noise have increased substantially in the previous three years, and that they are more common at night. See El-Bahr (1998:17).
54 Sengupta (1999) reports that residents of Calcutta slums also include moral pollution along with other forms of pollution.
55 2,009 respondents gave 2,135 answers. Because percentages are calculated based on the total number of respondents (and not number of answers), total percentages sometimes add up to more than 100%. Percentages for the localities and the genders are also calculated against the number of those who responded in each category.
56 Because multiple answers were possible—2,680 answers for 2,233 respondents, 33 not answering—answers cannot simply be totaled.
57 Again, multiple answers were possible—2,214 respondents gave 2,528 answers, of whom 5 said there was no water pollution.
58 Abkhas had the largest number of missing answers (16%).
59 Sengupta found essentially the same results in the Calcutta slums (*bastis*) he studied (1999:1293–94): " . . .the overwhelming majority in all the nine *basti* communities defined and perceived the environmental problems in terms of unhygienic physical surroundings and lack of basic services such as sanitation and water supply. Some have identified the deteriorating social environment as the major problem. Among the major social problems, drinking and hooliganism by local rowdies are the ones that concern most residents. The responses suggest that what the slum-dwellers want is not stricter pollution-control measures to preserve ecological balance and so on, but regular removal of accumulated garbage and adequate water supply and toilet facilities."
60 The Abt Associates study in 1982 noted that "garbage in the street" was the most frequently cited example of what people did not like about their neighborhood, followed by "flies and insects" and "overflowing sewers." The major sources of satisfaction were the social environment and adequate transportation. Abt study quoted in Steinberg (1990:126).
61 However, when we asked a related question, 77% of the Kafr al-Elow sample list air pollution as the biggest environmental problem. See Table 11 below.
62 People were offered a list of 11 possible forms of pollution, and asked to rank them. We have constructed this table by considering only the first choices. We explored various other ways of utilizing this data, and concluded that this was the most straightforward and did not distort the rest of the data.

Since people often did not complete the list it is impossible to construct an overall numerical score.

63 Not everyone answered: for instance only 95.9% in Kafr al-Elow gave a first choice.

64 If those in Abkhas who said there was no noise pollution are discarded, then the percentages of the remaining 250 were 38.8% for overcrowdedness, 35.2% for vehicles, 18% for microphones and 1.6% for factories and workshops.

65 This question allowed of multiple answers, so percentages can add to more than 100%. The percentages relate to the total sample. It should also be noted that the first four possibilities, which most people chose, were specified in the questionnaire (precoded, but also therefore closed-ended); the others were volunteered by respondents.

66 The sense of the question is that the government should be faulted for causing environmental pollution, rather than for failing to correct problems, but it is possible that some of those who gave this answer understood the second meaning.

67 In India it is fairly common to blame pollution problems on irresponsible private business and industry (see Gadgil and Guha 1995); this answer was essentially absent in Egypt.

68 In particular, those who answered that there was no other choice may have been thinking of themselves as well as of others.

69 In a 1997 study carried out by Environics International of Toronto, Canada, respondents in 24 countries, none in the Arab world, were asked which of these statements they agreed with: (1) Protecting the environment should be given priority, even at the risk of slowing down economic growth, or (2) Economic growth should be given priority, even if the environment suffers to some extent. This question is different than ours, so results cannot be directly compared. The fully developed countries were most likely to prefer the first option, with New Zealand on top, and such struggling countries as Nigeria and Ukraine were least likely to prefer that option. Nevertheless, on the strength of our results, it would seem likely that Egypt would fit in the middle range with such countries as Italy, India, China, Russia, or Spain, where opinion was pretty evenly split (47% to 53% preferred the first option). See Anderson and Smith (1997), prepared on the basis of the *Environmental Monitor*, Environics Ltd, Toronto, Canada.

70 Religious affiliation was not a factor in the choice of sample. Nearly all the people in the localities we chose seem to be Muslim. We recorded few if any Christians in our sample, and we have no information on any possible similarity or difference between the two religious traditions in this regard.

71 These two statements recorded the second and third highest "strong agreement." First was the proposition that "People have the right to clean air and water" (76.9%). These results were pretty even across the four localities.

72 In this question, multiple answers were possible; percentages here were calculated on the basis of those who answered, with "don't knows" omitted (n=2204).

73 Lane (1997) analyzes the role of television spots on health issues, while Schaufelberger (1995) gives an ethnographic view of television watching in general.

74 This contrast between men and women is widespread. See Heiberg and Øvensen (1993) and Giacaman et al. (1993).

75 These figures are much lower than those given by Tekçe et al. (1994:120–21), in a survey carried out in Manshiyet Nasser in 1984. For the two weeks preceding that survey, "a respiratory illness was reported for 49% of the children and diarrhea for 42%."

76 The relatively educated thus complain less about health problems while being more likely to think that pollution is the cause.

77 A study by Dr. Ahmed Sherif Hafez of the Faculty of Medicine at Ain Shams University (Hafez 1998) compared conditions in Helwan and Tenth of Ramadan City (a new city located in the desert about 45km east of Cairo), with the former considered to suffer more from air pollution. The study concluded that children in the Helwan area had lower body weight and height for their age, and suffered more from respiratory diseases, than those in Tenth of Ramadan. But in addition to the differential air pollution, there were also social class differences, as measured by father's education, which likewise favored Tenth of Ramadan.

78 The article by Bakr (1998) focuses on how lead pollution may contribute to mental retardation, more relevant for children.

79 In a 1997 survey in 24 countries conducted by Environics International Ltd. of Toronto, Canada, respondents were asked whether they felt that their health was affected by environmental problems. The people who said that their health was affected a great deal or a fair amount amounted to 94% in India, the highest score, 93% in China, 92% in Hungary, etc. Egypt was not included in the survey. (Anderson and Smith 1997:A15).

80 A sewage system was in fact installed several years later.

81 This sense of neglect and marginalization is the main theme of the recent book by Al-Hadini (1999), based on research in Arab al-Hosn in Matariyya and al-Munira al-Gharbiyya in Imbaba.

82 The EEAA was placed under the new Minister of State for Environmental Affairs in the summer of 1997, subsequent to our fieldwork.

83 Above we saw that the respondents in Abkhas were least likely to know of an NGO that was working to improve the environment, so this answer may be contradictory.

84 This is five times what the biggest spender in Manshiyet al-Sadat/Dar al-Salam reported (el-Ramly 1996:86).

85 For a similar situation in Morocco, see Navez-Bouchanine (1993).
86 Compare with the answers below on drivers maintaining their motors.
87 Our data suggest that the people in Abkhas think of the police and the courts rather than complaining to authorities. To reconcile this apparent anomaly requires a different kind of probing.
88 The role of these two newspapers in the 1995 elections is highlighted by Longuenesse (1997).
89 Ma'asara is described as a "poisoned neighborhood" (Abdel-Salam and Taha 1999).
90 The Helwan area is regularly singled out in the Egyptian media as the example of a heavily polluted area in which the population suffers.
91 Our field researchers observed a case where a complainant was misled by a local government official who deliberately sent her to the wrong place. This trick suggests some of the limits to complaining.
92 Although the numbers are much smaller, there were more people in Abkhas than elsewhere who saw crime as a problem—see Table 12.
93 The collection edited by Sandrine Gamblin (1997) provides a superb overview of the 1995 parliamentary elections. The article by Boutet in this collection specifically addresses the role of environmentally-oriented candidates, concluding that they are marginal. See also al-Karanshawy (1997) on the 1995 election in a Delta district, and Youssef and Longuenesse (1999) on Sayyida Zeinab.
94 These could be candidates for local office or for parliament. In the Egyptian electoral system for parliament, each constituency has two representatives, one of whom comes from the "workers and peasants" category and the other from an open category known as "other" (*fi'at*).
95 Both Mahgoub and Fath al-Bab were reelected in the fall 2000 elections.
96 Interview by a team member, October 1996. See also Nasr and Leila (1996).
97 See *Roz al-Yusif* for 28 January 1996. This lawyer often files such "public interest" cases.
98 The speaker is referring to a policy of relocating potentially polluting activities to the edge of the city. Al-Qattamiyya is in the desert on the eastern outskirts of Cairo.
99 M. Bell (1998:169–70) cites an example from an English village where male and female attitudes toward nature were strikingly different. We did not find such contrasts.
100 On the other hand, women were much more likely than men to say their household is connected to the water system (see Table 7).

Works Cited

Abdel Motaal, Doaa. 1997. "Reconstruction Development: Women at the Muqattam Settlement" in *The Zabbalin Community of Muqattam*. (*Cairo Papers in Social Science* 19[4]), Cairo: The American University in Cairo Press, pp. 59–110.

Abdel Salam, Dalia. 1997. "La capitale se met au vert" in *Al-Ahram Hebdo*, May 14, 1997, p. 4.

Abdel-Salam, Dalia. 1999. "Le nuage qui étouffe le Caire" and "La qualité de l'air en Egypte se dégrade sans cesse" in *Al-Ahram Hebdo*, Nov. 3, 1999, p. 7.

Abdel-Salam, Dalia and Ahmed Taha. 1999. "Al-Maasara: quartier empoisonné" in *Al-Ahram Hebdo*, September 22, 1999, p. 34.

Abt Associates. 1982. *Informal Housing in Egypt*. Cambridge, MA: ABT and Cairo: General Organization for Housing, Building, and Planning Research.

Abu-Lughod, Janet L. 1971. *Cairo: 1001 years of The City Victorious*. Princeton: Princeton University Press.

———. 1972. "Rural migration and politics in Egypt" in Richard Antoun and Iliya Harik, eds., *Rural Politics and Social Change in the Middle East*. Bloomington: Indiana University Press, pp. 315–334.

Abu-Taleb, Maher F. and Elias Salameh. 1994. "Environmental Management in Jordan: Problems and Recommendations" in *Environmental Conservation*, 21(1): 35–40.

Abu-Zahra, Nadia. 1997. *The Pure and the Powerful: Studies in Contemporary Muslim Society*. Reading UK: Garnet/Ithaca.

Al-Ahram Weekly. 1999. "Smoke gets in your eyes," December 16, 1999, p. 3.

American Chamber of Commerce. 2000. "Solid Waste Management in Egypt," *Sector Studies Series* 1. Cairo: American Chamber of Commerce.

Anderson, John, and Dita Smith. 1997. "Green, greener, greenest" in *Washington Post*, November 22, 1997, p. A15.

Arnaud, Jean-Luc. 1991. "Des jardins à la ville" in *Egypte-Monde Arabe* #8, pp. 87–105.

Assaad, Marie, and Nadra Garas. 1994. *Experiments in community development in a zabbaleen settlement*. (*Cairo Papers in Social Science* 16[4]), Cairo: The American University in Cairo Press.

Awad, Thabet Amin. 1999. "Percentage of Cairo's pollution is three times what is permissible" [*Nisib al talawuth bil-Qahira thalath amthal al-masmuh biha*] in *al-Ahram*, October 31, 1999, p. 22.

El-Bahr, Sahar. 1998. "Hello, emergency" in *Al-Ahram Weekly*, July 30, 1998, p. 17.

Bakr, Mahmoud. 1998. "Chains of molten lead" in *Al-Ahram Weekly*, March 12, 1998, p. 15.

Bakr, Mahmoud. 1999. "Lifting the veil of smog" [interview with Nadia Makram Ebeid] in *Al-Ahram Weekly*, Nov. 4, 1999, p. 5.

Baracat, Rafik. 1998. "Les ordures: fléau ou ressource?" in *Le Progrès Egyptien*, September 14, 1998, p. 4.

Baviskar, Amita. 1995. *In the Belly of the River: Tribal Conflicts over Development in the Narmada Valley*. Delhi, Oxford University Press.

Bayat, Asef. 1997. *Street Politics: Poor People's Movements in Iran*. New York: Columbia University Press.

Beck, Ulrich. 1992. "From industrial society to the risk society: questions of survival, social structure and ecological enlightenment" in *Theory, Culture and Society* 9:97–123.

Bell, Jennifer. 2000. "Egyptian Environmental Activists' Uphill Battle" in *Middle East Report*, #216, 30(3), 24–25.

Bell, Michael Meyerfeld. 1998. *An invitation to environmental sociology*. Thousand Oaks/London/New Delhi: Pine Forge Press.

Berkes, Fikret, ed. 1989. *Common Property Resources: Ecology and Community-Based Sustainable Development*. London: Belhaven Press.

Boutet, Annabelle. 1997. "Les écologistes en campagne: l'environnement comme nouvelle donne politique?" in Sandrine Gamblin, ed. *Contours et détours du politique en Egypte: les élections législatives de 1995*. Paris: L'Harmattan and Cairo: CEDEJ. pp. 151–63.

Briscoe, John. 1993. "When the cup is half full: improving water and sanitation services in the developing world" in *Environment* 35(4):7–37.

Bullard, Robert D. 1990. *Dumping in Dixie: Race, Class and Environmental quality*. Boulder: Westview.

Buttel, Frederick H. 1987. "New directions in environmental sociology" in *Annual Review of Sociology* 13:465–88.

Campbell, Tim. 1989. "Environmental dilemmas and the urban poor" in H. Jeffrey Leonard et al., *Environment and the Poor: Development Strategies for a Common Agenda*. New Brunswick: Transaction Books for the Overseas Development Council of Washington DC.

Čapek, Sheila M. 1993. "The 'environmental justice' frame: a conceptual discussion and an application" in *Social Problems* 40:5–24.

Carson, Rachel. 1962. *Silent Spring*. Boston: Houghton Mifflin.

El-Cherbini, Maha, 1997. "Tous reponsables, tous coupables" in *Al-Ahram Hebdo*, February 12, 1997, pp. 16–17.

Cole, Donald P. and Soraya Altorki. 1998. *Bedouins, Settlers, and Holiday-Makers: Egypt's Changing Northwest Coast.* Cairo: The American University in Cairo Press.

Douglas, Mary. 1970. *Purity and Danger: An Analysis of Concepts of Pollution and Taboo.* Harmondsworth: Penguin.

———. 1992. *Risk and Blame: Essays in Cultural Theory.* London and New York: Routledge.

Douglas, Mary, and Aaron Wildavsky. 1982. *Risk and Culture: An Essay on the Selection of Technological and Environmental Dangers.* Berkeley: University of California Press.

Edelstein, Michael R. 1988. *Contaminated Communities: The Social and Psychological Impacts of Residential Toxic Exposure.* Boulder: Westview Press.

Egyptian Gazette. 1998. "Jordan bails officials in water pollution row," September 22, 1998 (Reuters dispatch).

Egyptian Gazette. 1999. "Cabinet addresses smoggy clouds, earmarks LE 275m. for garbage recycling," October 31, 1999, p. 1.

Engelman, Robert, and Pamela LeRoy. 1993. *Sustaining Water: Population and the Future of Renewable Water Supplies.* Washington DC: Population Action International.

Euroconsult/Darwish Consulting Engineers. 1992. *Environmental Profile, Fayoum Governorate, Egypt, Background Study.* Netherlands Directorate for International Cooperation, June.

Fakhouri, Hani. 1987. *Kafr El-Elow: Continuity and Change in an Egyptian Community.* 2nd ed., with an additional chapter, pp. 127–56. Prospect Heights IL: Waveland Press. (orig. *Kafr El-Elow: An Egyptian Village in Transition.* New York: Holt, Rinehart, Winston, 1972.)

Farrag, Eftetan. 1995. "Working children in Cairo: case studies" in Elizabeth Warnock Fernea, ed., *Children in the Muslim Middle East.* Austin: University of Texas Press, pp. 239-49.

Feeny, David, Fikret Berkes, Bonnie J. McCay, and James Acheson. 1990. "The tragedy of the commons: twenty-two years later" in *Human Ecology* 18(1):1–19.

Fitchen, Janet M. 1988. "Anthropology and environmental problems in the U.S.: the case of groundwater contamination" in *Practicing Anthropology* 10(3–4):18–20.

Fowlkes, Martha R., and Patricia Y. Miller. 1987. "Chemicals and community at Love Canal" in B.B. Johnson and V.T. Covello, eds, *The Social and Cultural Construction of Risk: Essays on Risk Selection and Perception.* Dordrecht/Boston: D. Reidel, pp. 55–78.

Gad, Mona Ahmed. 1998. "The reasons for environmental pollution in Sayyida Zeinab and its effects on human health and the role of participation in ending environmental problems" in Nicholas S. Hopkins, Sohair R. Mehanna, and Salah el-Hagar, *Social Response to Environmental Change and Pollution in Egypt*. Cairo, Social Research Center, pp. 219–30.

Gadgil, Madhav, and Ramachandra Guha. 1995. *Ecology and Equity: The Use and Abuse of Nature in Contemporary India.* London and New York: Routledge for UNRISD.

Gamblin, Sandrine, ed. 1997. *Contours et détours du politique en Egypte: les élections législatives de 1995.* Paris: L'Harmattan and Cairo: CEDEJ.

Geertz, Clifford. 1973. *The Interpretation of Cultures*. New York: Basic Books.

Gheith, Chahinaz. 2000. "Coup de balai non gouvernemental: Depuis 1993, le Bureau arabe pour les jeunes et l'environnement (ONG) multiplie les initiatives" in *Al-Ahram Hebdo*, March 15, 2000, p. 27.

Giacaman, Rita, Camilla Stoltenberg, and Lars Weiseth. 1993. "Health," in Marianne Heiberg and Geir Ovensen, *Palestinian Society in Gaza, West Bank and Arab Jerusalem: A Survey of Living Conditions.* Oslo: FAVO, pp. 99–130.

el-Gohary, Fatma. 1994. "Comparative environmental risks in Cairo: water pollution problems" in *Comparing Environmental Health Risks in Cairo, Egypt.* 3 vols. Cairo: USAID. Vol. 3, Annex H.

el-Gohary, Fatma A., Fayza A. Nasr, and S. El-Hawaary. 1998. "Performance assessment of a wastewater treatment plant producing effluent for irrigation in Egypt" in *The Environmentalist* 18:87–93.

Gomaa, Salwa Sharawi. 1992. "Environmental Politics in Egypt" in Nicholas S. Hopkins, ed., *Environmental Challenges in Egypt and the World.* (*Cairo Papers in Social Science* 15[4]) Cairo: The American University in Cairo Press, pp. 24–40.

Gomaa, Salwa Sharawi. 1997. *Environmental Policy Making in Egypt.* Gainesville: University Press of Florida and Cairo: The American University in Cairo Press.

Guha, Ramachandra. 1997. "The environmentalism of the poor" in Richard G. Fox and Orin Starn, eds., *Between Resistance and Revolution: Cultural Politics and Social Politics.* New Brunswick NJ: Rutgers University Press. pp. 17–39.

Al-Hadini, Amani Mesa'ud. 1999. *Al-Muhamishun w'al-siyasa fi Misr* (The marginalized and politics in Egypt). Cairo, Center for Political and Strategic Studies.

Hafez, Ahmed Sherif. 1998. *Impact of Outdoor Air Pollution on Anthropometry of Preschool Children in Two Urban Areas (a Preliminary Study).* Cairo: National Population Council and Ain Shams University, Faculty of Medicine, Environment and Community Affairs.

El-Hakim, S.M. 1981. "Social policy and the environmental issue: the case of Cairo's domestic waste" in *Die Dritte Welt,* 9(1–2):100–107.

Hannigan, John A. 1995. *Environmental Sociology: A Social Constructionist Perspective.* London: Routledge.

Hardin, Garrett. 1968. "The tragedy of the commons" in *Science* 162:1243–48.

Hardoy, Jorge E., Sandy Cairncross, and David Satterthwaite, eds. 1990. *The Poor Die Young: Housing and Health in Third World Cities.* London: Earthscan.

Heiberg, Marianne, and Geir Øvensen et al. 1993. *Palestinian Society in Gaza, West Bank and Arab Jerusalem: A Survey of Living Conditions.* Oslo: FAVO.

Hobbs, Joseph J. 1995. *Mount Sinai.* Austin: University of Texas Press, and Cairo: The American University in Cairo Press.

Hoodfar, Homa. 1998. *Between Marriage and the Market: Intimate Politics and Survival in Cairo.* Cairo: The American University in Cairo Press.

Hopkins, Nicholas S. et al. 1988. *Participation and Community in the Egyptian New Lands: the Case of South Tahrir.* (*Cairo Papers in Social Science* 11[1]), Cairo: The American University in Cairo Press.

Hopkins, Nicholas S. et al. 1995. "Pollution and people in Cairo" in *Environmental Threats in Egypt: Perceptions and Actions,* ed. Salwa Sharawi Gomaa. (*Cairo Papers in Social Science* 17[4]), Cairo: The American University in Cairo Press, pp. 3–28.

Hopkins, Nicholas S., and Sohair Mehanna. 1997. "Pollution, popular perceptions and grassroots environmental activism" in *Middle East Report* #202, 27[1], pp. 21–25, winter.

Hopkins, Nicholas S., and Sohair Mehanna. 2000. "Social action against everyday pollution in Egypt" in *Human Organization* 59(2): 245–54.

Hopkins, Nicholas S., Sohair R. Mehanna, and Salah el-Hagar. 1998. *Social Response to Environmental Change and Pollution in Egypt.* Cairo: Social Research Center for the International Development Research Centre, Canada.

Ibrahim, Fouad N. 1996. *Ägypten: eine geographische Landeskunde.* Unter Mitarb. von Barbara Ibrahim. Darmstadt: Wissenschaftliche Buchgesellschaft.

Ibrahim, Saad Eddin. 1995. "State, women, and civil society: an evaluation of Egypt's population policy" in Carla Makhlouf Obermeyer, ed., *Family,*

Gender, and Population in the Middle East: Policies in Context. Cairo: The American University in Cairo Press, pp. 57–79.
Ibrahim, Saad Eddin et al. 1997. *An Assessment of Grass Roots Participation in the Development of Egypt.* (*Cairo Papers in Social Science* 19[3]), Cairo: The American University in Cairo Press.
Ireton, François. 1988. "'Du limon au beton': l'urbanisation spontanée à Bulaq el-Dakrur" in *Bulletin du CEDEJ* #24, pp. 121–50.
Ireton, François. 1998. "The evolution of agrarian structures in Egypt: regional patterns of change in farm size" in Nicholas S. Hopkins and Kirsten Westergaard, eds., *Directions of Change in Rural Egypt.* Cairo: American University in Cairo Press. pp. 41–65.
Ismail, Salwa. 1996. "The politics of space in urban Cairo: informal communities and the state" in *Arab Studies Journal* 4(2):119–32.
Jamison, Andrew, Ron Eyerman, and Jacqueline Cramer. 1990. *The Making of the New Environmental Consciousness: A Comparative Study of the Environmental Movements in Sweden, Denmark and the Netherlands.* Edinburgh: Edinburgh University Press.
Johnson, Branden, and Vincent Covello, eds. 1987. *The Social and Cultural Construction of Risk: Essays on Risk Selection and Perception.* Dordrecht/Boston: D. Reidel.
El-Kadi, Galila. 1987. *L'urbanisation spontanée au Caire.* Tours: URBAMA, Fascicule de Recherches no. 18.
Kadri, Samir. "Cairo smog not local problem, part of global warming" in *Egyptian Gazette* December 10, 2000, p. 3.
Kamel, Khaled. 1994. "A tale of two factories" in Salwa Sharawi Gomaa, ed., *Environmental Threats in Egypt: Perceptions and Actions.* (*Cairo Papers in Social Science* 17[4]), Cairo: The American University in Cairo Press, pp. 29–39.
el-Karanshawy, Samer. 1997. *Class, Family and Power in an Egyptian Village.* (*Cairo Papers in Social Science* 20[1]), Cairo: The American University in Cairo Press, pp. 1–74.
Kasperson, Roger E. et al. 1988. "The social amplification of risk: a conceptual framework" in *Risk Analysis* 8(2):177–187.
El Katsha, Samiha, A. Younis, O. El-Sebaie, and A. Hussein. 1989. *Women, Water, and Sanitation: Household Water Use in Two Egyptian Villages. (Cairo Papers in Social Science* 12[2]), Cairo: The American University in Cairo Press.
Kempton, Willett, James S. Boster, and Jennifer A. Hartley. 1995. *Environmental Values in American Culture.* Cambridge MA: The MIT Press.

Khalil, Ashraf. 1999. "Smog spurs cleanup effort" in *Business Monthly*, December 1999, 15 (12): 36–37.
Kharoufi, Mostafa. 1991. "Du petit au grand espace urbain: le commerce des fruits et légumes à Dar al-Salam" in *Egypte-Monde Arabe* #5, pp. 81–96.
Kharoufi, Mostafa. 1995. "Société et espace dans un quartier du Caire (Dar el-Salam): Secteur 'informel' et intégration urbaine" in *Les Cahiers de l'URBAMA* no. 11, pp. 57–91.
Khattab, Hind. 1992. *The Silent Endurance: Social Conditions of Women's Reproductive Health in Rural Egypt.* Cairo: UNICEF and the Population Council.
Kottak, Conrad P. 1992. *Assault on Paradise: Social Change in a Brazilian Village.* 2nd ed. New York: McGraw-Hill.
Kottak, Conrad. 1999. "The new ecological anthropology" in *American Anthropologist* 101(1):23–35.
Lane, Sandra. 1997. "Television minidramas: social marketing and evaluation in Egypt" in *Medical Anthropology Quarterly* 11(2):164–82.
Levine, Adeline Gordon. 1982. *Love Canal: Science, Politics, and People.* Lexington MA: Lexington Books/D.C. Heath.
Lindén, Anna-Lisa, ed. 1997. *Thinking, Saying, Doing: Sociological Perspectives on Environmental Behaviour.* Lund: Department of Sociology, Lund University, Research Report #2.
Longuenesse, Elisabeth. 1997. "Logiques d'appartenances et dynamique électorale dans une banlieue ouvrière: le cas de la circonscription 25 à Helwan" in Gamblin, Sandrine, ed, *Contours et détours du politique en Egypte: les élections législatives de 1995.* Paris: L'Harmattan and Cairo: CEDEJ. pp. 229–65.
Lovei, Magda. 1998. *Phasing Out Lead from Gasoline: Worldwide Experience and Policy Implications.* Washington: World Bank Technical Paper no. 397, Pollution Management Series.
Loza, Sarah F. 1992. "Urban street food vendors: case study from Egypt" *Informal Sector in Egypt*, ed. Nicholas S. Hopkins. (*Cairo Papers in Social Science* 14[4]), Cairo: The American University in Cairo Press. pp. 46–52.
McCay, Bonnie J., and James M. Acheson, eds. 1987. *The Question of the Commons: The Culture and Ecology of Communal Resources.* Tucson: University of Arizona Press.
McGurty, Eileen Maura. 1997. "From NIMBY to Civil Rights: The Origins of the Environmental Justice Movement" in *Environmental History* 2(3):301–23.
Mehanna, Sohair, Nicholas S. Hopkins, and Bahgat Abdelmaksoud. 1994. "Farmers and Merchants: Background to Structural Adjustment." (*Cairo*

Papers in Social Science 17[2]), Cairo: The American University in Cairo Press.

Meyer, Günter. 1987. "Waste recycling as a livelihood in the informal sector: the example of refuse collectors in Cairo" in *Applied Geography and Development* 30:78–94.

Miller, G. Tyler. 1998. *Living in the Environment.* 10th ed. Belmont CA: Wadsworth.

Ministry of State for Environmental Affairs (MSEA). 2000. "The Environmental Profile of Egypt." Cairo.

Montasser, Salah. 1999. "An opinion: The voice of the people," [*Mugarrad ra'y: sawt al-sha'b]* in *al-Ahram*, November 1, 1999, p. 11.

Nadim, Asaad et al. 1980. *Living Without Water.* (*Cairo Papers in Social Science* 3[3]), Cairo: The American University in Cairo Press.

Nadim, Nawal al-Messiri. 1985. "Family relationships in a harah in Cairo" in *Arab Society: Social Science Perspectives*, ed. Nicholas S. Hopkins and Saad Eddin Ibrahim. Cairo: American University in Cairo Press, pp. 212–22.

Nasr, Sherine. 2000. "Living under the cloud" in *Al-Ahram Weekly*, November 9, 2000, p. 2.

Nasr, Sherine, and Reem Leila. 1996. "Lawyer battles cement plants" in *Al-Ahram Weekly*, September 12, 1996, p. 2.

Nasr, Sherine, and Mahmoud Bakr. 1997a. "Clouds that won't go away: Breathlessly chasing the environment" in *Al-Ahram Weekly*, June 5, 1997, p. 13.

Navez-Bouchanine, Françoise. 1993. "Propreté et appropriation de l'espace: déchets et pratiques habitantes dans les grandes villes marocaines" in *Peuples Méditerranéens*, no. 62–63, pp. 165–82.

Newman, Penny. 1992. "Killing legally with toxic waste: women and the environment in the United States" in *Development Dialogue* (1–2):50–70.

Nielsen, Hans-Chr. Korsholm. 1998. "Men of authority—documents of authority: notes on customary law in Upper Egypt." In *Directions of Change in Rural Egypt.* Nicholas S. Hopkins and Kirsten Westergaard, eds. pp. 357–70. Cairo: The American University in Cairo Press.

van Nieuwenhuizje, C.A.O., M. Fathalla al-Khatib, and Adel Azer. 1985. *The Poor Man's Model of Development.* Leiden: Brill.

Ostrom, Elinor. 1990. *Governing the Commons: The Evolution of Institutions for Collective Action.* Cambridge UK: Cambridge University Press.

Othman, Amira Ahmed Fouad, "The Islamic cult of saints and patron-client relationships (the case of Al-Sayyida Zeinab)," MA thesis,

Department of Sociology-Anthropology, American University in Cairo, 1997.

O'Toole, Laurence J., Janet A. Phoenix, and M. Walid Gamaleldin. 1996. *Institutional Assessment for Lead Exposure Abatement and Reduction in Cairo*. Environmental Health Project, Activity Report no. 31. Washington DC: USAID.

Petterson, John S. 1988. "The reality of perception: demonstrable effects of perceived risk in Goiania, Brazil" in *Practicing Anthropology* 10(3–4):8–9, 12.

PRIDE (Chemonics International). 1994. *Comparing Environmental Health Risks in Cairo, Egypt*. 3 vols. Cairo: USAID.

Proceedings of the Conference on Common Property Resource Management. 1985. Washington: National Academy Press.

Putnam, Robert D. 1995. "Bowling alone: America's declining social capital" in *Journal of Democracy* 6(1):65–78.

Raafat, Samir W. 1994. *Maadi 1904–1962: Society and History in a Cairo Suburb*. Cairo: Palm Press.

el-Ramly, Eman Hassan. 1996. "Women's perception of environmental change in Cairo." MA thesis, Department of Sociology-Anthropology, American University in Cairo, 1996.

Sachs, Aaron. 1995. *Eco-Justice: Linking Human Rights and the Environment.* Washington: Worldwatch Institute, Paper #127.

Said, Rushdi. 1993. *The River Nile: Geology, Hydrology and Utilization.* Oxford: Pergamon Press.

Said, Rushdi. 2000. "Saved, or lost forever [Earth Day 2000]" in *Al-Ahram Weekly*, April 20, 2000, p. 11.

Schaufelberger, Esther. 1995. "Eating *tami'a* as if it were *kabab*: An Egyptian television serials response to economic transformation" in *The Metropolitan Food System of Cairo*, Jörg Gertel, Saarbrücken, ed. Verlag für Entwicklungspolitik, Freiburger Studien zur Geographischen Entwicklungsforschung 13, pp. 161–83.

Sengupta, Chandan. 1999. "Dynamics of community environmental management in Howrah slums" in *Economic and Political Weekly* 34(21):1292–1296 (May 22, 1999).

Singerman, Diane. 1997. *Avenues of Participation: Family, Politics, and Networks in Urban Quarters of Cairo*. Cairo: The American University in Cairo Press.

Steinberg, Florian. 1990. "Cairo: informal land development and the challenge for the future" in Paul Baróss and Jan van der Linden, eds., *The Transformation of Land Supply Systems in Third World Cities.* Aldershot (UK): Avebury for Gower, pp. 111–32.

Tadros, Mariz. 1996. "Sewage siege" in *Al-Ahram Weekly*, June 6, 1996, p. 14.

———. 1998. "Another Helwan?" in *Al-Ahram Weekly*, March 12, 1998, p. 15.

———. 1999a. "Without silver linings" in *Al-Ahram Weekly*, November 4, 1999, p. 1.

———.. 1999b. "The plot thickens" in *Al-Ahram Weekly*, November 4, 1999, p. 5.

Tekçe, Belgin, Linda Oldham, and Frederic Shorter. 1994. *A Place to Live: Families and Child Health in a Cairo Neighborhood.* Cairo: The American University in Cairo Press.

Tewfik, Inas. 1996. "Community participation and environmental change: mobilization in a Cairo neighborhood." MA thesis, Department of Sociology-Anthropology, American University in Cairo.

Tewfik, Inas. 1997. "Mobilization in a Cairo Neighborhood: Community Participation and Environmental Change" in *Middle East Report* #202, pp. 26–27, winter.

Thomas, W. I. 1923. *The Unadjusted Girl.* New York: Little, Brown.

Tolba, Mustafa. 1993. "Environment and Development," lecture given at the American University in Cairo for Earth Day, April 24, 1993.

Volpi, Elena. 1997. "Community organization and development" *The Zabbalin Community of Muqattam.* (*Cairo Papers in Social Science* 19[4]), Cairo: The American University in Cairo Press, pp. 5–58.

Wapner, Paul. 1996. *Environmental Activism and World Civic Politics.* Albany NY: State University of New York Press.

Watts, Susan, and Samiha El Katsha. 1995. "Changing environmental conditions in the Nile delta: health and policy implications with special reference to schistosomiasis" in *International Journal of Environmental Health Research* 5:197–212.

———. 1997. "Irrigation, farming and schistosomiasis: a case study in the Nile Delta" in *International Journal of Environmental Health Research* 7:101–13.

White, Gilbert F. 1988. "Paths to risk analysis" in *Risk Analysis* 8(2):171–75.

Wikan, Unni. 1995. "Sustainable development in the mega-city: can the concept be made applicable?" in *Current Anthropology* 36(4):635–55.

Wildavsky, Aaron, and Karl Dake. 1990. "Theories of risk perception: who fears what and why?" in *Daedalus* 119(4):41–60.

Wolfe, Amy K. 1988. "Environmental risk and anthropology" in *Practising Anthropology* 10(3–4):4.

World Health Organization and United Nations Environment Programme. 1992. *Urban Air Pollution in Megacities of the World.* Oxford: Blackwell and WHO/UNEP.

World Resources Institute. 1996. *A Guide to the Global Environment: The Urban Environment.* New York: Oxford University Press.

Youssef, Abdel Maseeh Felly, and Elisabeth Longuenesse. 1999. "Affaires et politiques au Caire: l'exemple du quartier de Sayyeda Zaynab" in *Monde Arabe Maghreb-Machrek*, #166, pp. 53–69.

Zayed, Ahmed. 1998. "Culture and Mediation of Power in an Egyptian Village" in *Directions of Change in Rural Egypt.* Nicholas S. Hopkins and Kirsten Westergaard, eds. Cairo: American University in Cairo Press, pp. 371–88.

Zonabend, Françoise. 1993. *The Nuclear Peninsula.* Cambridge: Cambridge University Press (Maison des Sciences de l'Homme).

Index